THE COMMONWEALTH AND INTERNATIONAL
Joint Chairmen of the Honorary Editorial Advisory
SIR ROBERT ROBINSON, O.M., F.R.S., LONDON
DEAN ATHELSTAN SPILHAUS, MINNESOTA

GEOLOGY DIVISION
General Editor: F. H. T. RHODES

GEOLOGICAL MAPS

GEOLOGICAL MAPS

by BRIAN SIMPSON, M.Sc. (Liverpool)
Associate of the Institution of Civil Engineers, F.G.S.

SENIOR LECTURER IN GEOLOGY, UNIVERSITY COLLEGE, SWANSEA
CHIEF EXAMINER IN GEOLOGY AT ADVANCED LEVEL G.C.E. FOR THE
WELSH JOINT EDUCATION COMMITTEE

PERGAMON PRESS

OXFORD · NEW YORK
TORONTO · SYDNEY · BRAUNSCHWEIG

Pergamon Press Ltd., Headington Hill Hall, Oxford
Pergamon Press Inc., Maxwell House, Fairview Park, Elmsford, New York 10523
Pergamon of Canada Ltd., 207 Queen's Quay West, Toronto 1
Pergamon Press (Aust.) Pty. Ltd., 19a Boundary Street,
Rushcutters Bay, N.S.W. 2011, Australia
Vieweg & Sohn GmbH, Burgplatz 1, Braunschweig

Copyright © 1968 Brian Simpson

All Rights Reserved. No part of this publication may be reproduced, stored in a retrieval system, or transmitted, in any form or by any means, electronic, mechanical, photocopying, recording or otherwise, without the prior permission of Pergamon Press Ltd.

First edition 1968
Reprinted 1970
Library of Congress Catalog Card No. 67-31507

*Filmset by The European Printing Corporation Limited, Dublin, Ireland
Printed in Great Britain by A. Wheaton & Co., Exeter*

This book is sold subject to the condition
that it shall not, by way of trade, be lent,
resold, hired out, or otherwise disposed
of without the publisher's consent,
in any form of binding or cover
other than that in which
it is published.
08 012780 0 (flexicover)
08 012781 9 (hard case)

CONTENTS

Preface		vii
Key for Rock Symbols		viii
PART 1.	Topographic maps	1
	Simple land forms depicted by contours	1
	True and apparent dip and strike of beds of rock	3
	The effect of the dip of a stratum on its outcrop	3
	Methods for the determination of the dip and strike of a rock succession	5
	To determine the dip and strike of a rock series from outcrops on a geological map	7
	Outcrop trend and form in relation to topography	13
PART 2.	The determination of the thickness of a bed of rock	14
PART 3.	Folds	16
	Types of fold and their recognition on maps	16
PART 4.	Faults	26
	A. Descriptive terminology of faults	26
	B. The broad classification of faults	27
	C. The effect of faults on the outcrops of beds	33
	D. Two practical examples of the effect of faulting	33
	The determination of the vertical throw of a dip fault	33
PART 5.	Unconformities	49
	To determine the junction of the beds of a lower series with the base of the upper series in an unconformity	54
PART 6.	Outliers and inliers	55
PART 7.	To complete the outcrops of beds from partial outcrops	58
	To plot rock outcrops from borehole records	58
	The angle of dip in exaggerated vertical scale in geological sections	68
APPENDIX II.	Some graphical methods involving dip and strike problems	69

APPENDIX III. Relationship between true and apparent dip — 75

APPENDIX IV. The solution of dip and apparent dip problems using the stereographic projection — 75

APPENDIX V. Miscellaneous examples — 82

BIBLIOGRAPHY — 95

INDEX — 97

PREFACE

This book is intended for the use of students studying geology for the first time, particularly those entering the Advanced Level and First Year University examinations. It is not in any way an exhaustive treatise on geological maps, but is meant to serve as an introduction to their interpretation and solution. The book should not be regarded as an end in itself, nor should the somewhat mechanical manner of this early treatment be more than a means of developing the capacity for the three-dimensional viewing of a geological map and an appreciation of the patterns developed in rock relationships. Diagram maps must lead very quickly to the study of geological maps of specific areas of country: the ideal maps for such study are those produced by the Geological Surveys of such countries as Great Britain, Australia and the Americas.

My thanks are due to Professor F. H. T. Rhodes and to Dr. R. L. Austin for reading the manuscript and making many helpful suggestions. I would also thank Mrs. Greir Lewis for her help in the preparation of some of the diagrams and the Secretarial staff of University College, Swansea, for their help in typing the manuscript. I owe a particular debt of thanks to Mr. H. McKee who has prepared the index, and to Dr. K. G. Stagg who has read the proofs and helped in correcting them.

KEY FOR ROCK SYMBOLS

LIMESTONE		FINE SANDSTONE		MARL
LIMESTONE OR OOLITIC LIMESTONE		COARSE SANDSTONE		
SHALE		COARSE SANDSTONE		
MUDSTONE		CONGLOMERATE		
DYKE ROCKS E.G. DOLERITES		IGNEOUS E.G. GRANITE		

N.B. THE SYMBOLS REFER TO THOSE DIAGRAMS WHERE NO KEY HAS BEEN INCLUDED: OTHER KEYS ACCOMPANY THE MAPS TO WHICH THEY REFER.

PART 1

TOPOGRAPHIC MAPS

The basis for the compilation of a geological map is usually a topographic map on which diversity of land forms is expressed by means of contour lines, the latter being lines which join all points of equal height above mean sea level. The common scale used for geological surveying map work is 1/10,560 or 6 inches to 1 mile; for greater detail the 1/2500 or 25·344 inches to 1 mile is used. The published, coloured, sheets of the Geological Survey of Britain are on a scale of 1/63,360 or 1 inch to 1 mile.

The most useful map scale for studying the representation of land forms is generally 1 inch to 1 mile, since on this scale the relative closeness of contours gives an immediate visual impression, whereas with larger scales there is so great a distance between the contours that the impression is less marked.

SIMPLE LAND FORMS DEPICTED BY CONTOURS

It is important to appreciate the manner in which topographic diversity is expressed on maps so that the configuration of the land can be visualised quickly from the map or the land form sketched intelligently from a specific view. This is important to geologists since, in very great measure, the diversity of land form is the expression of the foundation geology of an area.

The usual method of expressing land forms is by contour lines and a few basic concepts of this method are expressed in Figs. 1A–G.

The contour line is a way of showing, on a map, the position of all points of like elevation above ordnance datum (expressed as O.D.) so that the 800 ft contour, for example, would be the line drawn on a map to indicate the locus of all points which were 800 ft above ordnance datum. Broadly the contours on a map show the distribution of high and low ground and indicate variations in gradient. The gradient for the same degree of separation of contours would be different on maps of different scales.

Figure 1A shows evenly spaced contours from 300 ft O.D. to 800 ft O.D. and the section along *X-Y* shows an even gradient and relatively gentle slope.

Figure 1B shows the closer spacing of the contours between 800 ft O.D. and 300 ft O.D. indicating a change of slope of the ground between these contours.

Figure 1C shows an abrupt change of slope at the 800 ft contour: the great closeness of the contours between 800 ft O.D. and 300 ft O.D. indicates a steep escarpment whilst the wide separation of the 300 ft and 200 ft contours reflects a gently dipping plane.

Figure 1D shows a sloping upland plain passing by a gentle escarpment into a valley slope.

Figure 1E shows a hill and valley landscape. Notice that the valley contours have the point of their V outline pointing towards the highest value contours whilst in the spurs, the point of the V is towards contours of lower value: note also the disposition of the contours in respect of slope as seen in the sections.

Figure 1F depicts a rounded hill; the upper part is shown by closed contours. Note that the western slope is more gentle than the eastern, as indicated by the contours.

Figure 1G shows a number of common topographic forms on a single map. The essential needs of the geologist are to be able to read a map accurately and interpret quickly the nature of a terrain. In order to become proficient in this field the student should study the 1 inch to 1 mile Ordnance Survey Maps, not only in the laboratory, but also in the field.

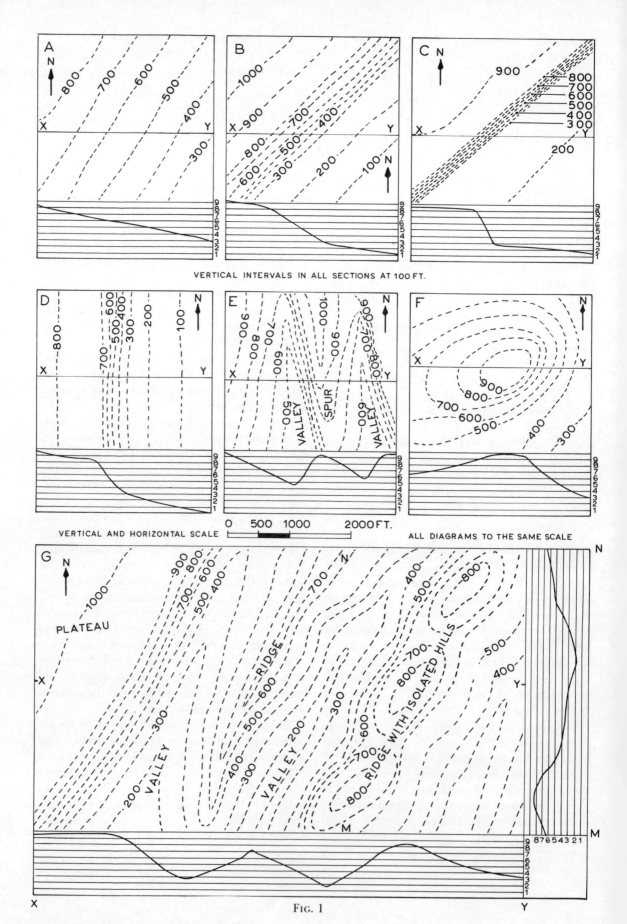

Fig. 1

TRUE AND APPARENT DIP AND STRIKE OF BEDS OF ROCK

As the above terms will recur frequently in the text they are defined at once, but a more extensive examination of these will occur later.

Most sedimentary rocks, many metamorphic rocks and igneous sills and dykes, occur as sheet-like bodies. When these "sheets" or beds are inclined to the horizontal they are said to *dip*. Figure 2 shows a bed of sandstone inclined at an angle $B\hat{A}C$ to the horizontal AC: the angle $B\hat{A}C$, which is the maximum inclination to the horizontal, is the angle of full or true dip. With the exception of the direction at right angles to the direction of true dip,

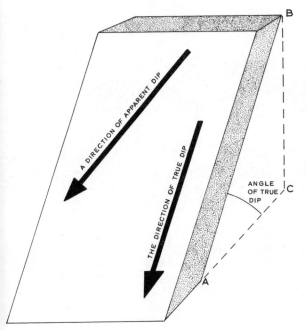

TRUE AND APPARENT DIP

Fig. 2

which is called the *strike* of the bed, every direction across the bed has some dip. The dip in any direction other than the true dip is called an *apparent dip*: such dips are always smaller than the true dip. Along the strike of the beds there is no dip, and the beds appear horizontal. For the relation between true and apparent dip see Appendix III.

THE EFFECT OF THE DIP OF A STRATUM ON ITS OUTCROP

Two features influence the appearance of the outcrop of a bed of rock; they are its dip and the form of the topography.

Figure 3A shows the outcrop of a bed of sandstone which is horizontal, a fact which is demonstrated in the section drawn along *X-Y*. It is apparent from the map that a bed which is horizontal will have outcrops, and hence bedding planes, which trend parallel to the topographic contours. That the slope of the country influences the width of outcrop is seen from a consideration of the outcrop widths *ab* and *AB*: in the former the slope of the surface is steeper than in the latter and its projection on the map is, therefore, the narrower of the two outcrops.

In Fig. 3C the dark bands are outcrops of vertical sheets of rock which project as straight bands on the map uninfluenced by any topographic diversity. The section along *X-Y* shows these vertical dykes and demonstrates that their thickness *in the section* varies with the angle made by the dyke with the section. At *D* the section is almost at right angles to the dyke which, therefore, almost shows its true thickness: at *A* and *B* the width of the dyke in the section is much greater than its true thickness since dyke and section are nearing parallelism.

In Fig. 3B the bed of sandstone has a dip towards the west. This can be established by considering the outcrop at *e*, *f* and *g* which lie at 500 ft, 400 ft and 300 ft O.D. respectively so that the slope of the bed is from *e* to *g*: this same kind of reasoning could be applied on the *upper surface* of the sandstone noting the fall from *d* to *h* or *d'* to *h'*. The importance of strike lines is discussed later; they are introduced here in order to explain the inclined nature of the beds. The true strike of the bed can be obtained by joining *ff'* and *gg'* on the lower surface of the sandstone or *dd'* or *h'h* on the upper surface, thus, in effect, drawing straight line contours on the surfaces of the beds. The true dip will be at right angles to the strike and, since between the strike line *ff'* and *gg'* there is a fall of 100 ft in a distance of 500 ft the gradient is 1 in 5, or the dip 11°20′. The figures in square brackets refer

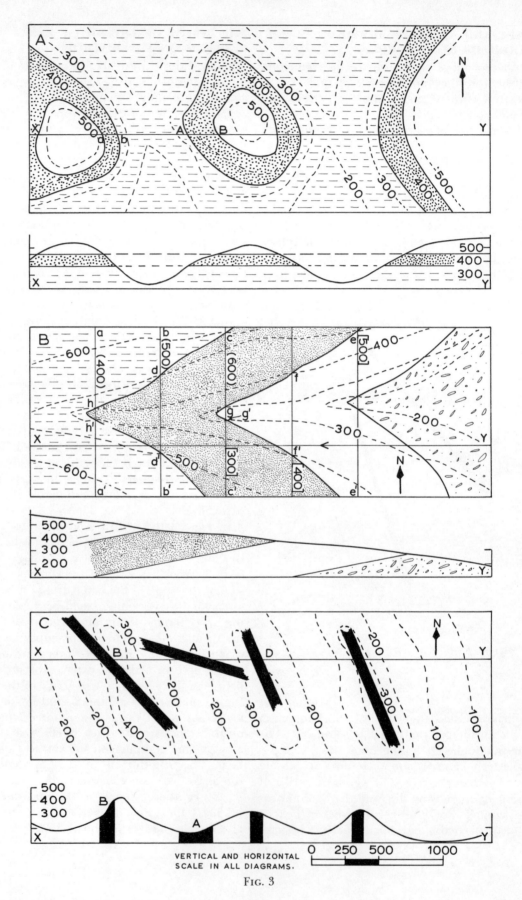

Fig. 3

to the base of the bed. Those in curved brackets refer to the upper surface.

In this case the effect of the interrelation between the dipping bed of sandstone and the topography is to develop a sinuosity in the outcrop: in this simple example there is a characteristic V pattern which, in a valley, becomes more acute and narrower between the sides of the V as the dip increases and the valley narrows. This is illustrated in Fig. 4, where the same bed with the same dip presents increasing diversity of outcrop as the complexity of the topography increases.

METHODS FOR THE DETERMINATION OF THE DIP AND STRIKE OF A ROCK SUCCESSION

1. *By Drawing Strike Lines*

Figure 5 shows the outcrops of a series of beds in an area of irregular topography: since the outcrops of the rocks cross the contours and are sinuous in their course they must be inclined to the horizontal, or more shortly — they have a dip.

The direction of strike may be determined in the manner set out below.

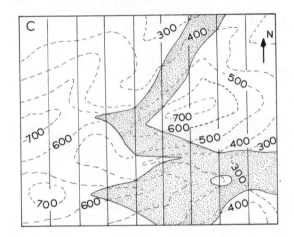

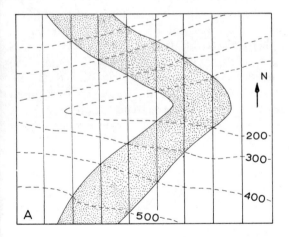

THE BED OF SANDSTONE IN THE THREE MAPS A, B AND C (ALL DRAWN TO THE SAME SCALE) HAS THE SAME THICKNESS AND THE SAME DIRECTION AND AMOUNT OF DIP. THE INCREASING COMPLEXITY OF THE OUTCROPS IN A, B AND C IS ENTIRELY THE RESULT OF TOPOGRAPHIC VARIATION.

MAPS TO ILLUSTRATE THE VARIATION IN THE FORM OF OUTCROP RESULTING FROM THE COMPLEXITY OF THE TOPOGRAPHY.

FIG. 4

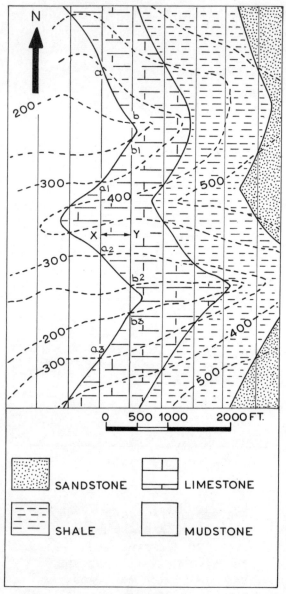

Fig. 5

Consider the junction of the mudstone and the limestone at the points a, a_1, a_2 and a_3: these points all lie at 300 ft O.D. where the outcrop of rock and land surface are at the same elevation. A line joining the four points is a strike line or stratum contour; it is, in fact, a contour line on a plane surface and is therefore straight. A second strike line drawn through b, b_1, b_2 and b_3 is parallel to the first and is the 200 ft O.D. level on the lower surface of the limestone. The dip of the lower surface of the limestone, which is the same as the upper surface of the mudstone, may now be determined: X lies on the 300 ft strike line and Y on the 200 ft strike line; therefore the surface of the bed falls 100 ft from X to Y, which may be shown by measurement from the map to be a horizontal distance of 500 ft. Since the surface falls in elevation eastwards and by definition the dip is at right angles to the strike, the direction of dip is due east. The dip of the lower surface of the limestone is 1 in 5 expressed as a gradient or, as an angle, 11°20′.

Strike lines may now be drawn for each bed through points of equal elevation on the same surface of a bed. On the map now being considered the strike lines are equidistant over the whole map and the same group of strike lines (but with different values) may be used for all the beds. This is a consequence of the beds being even hundreds of feet in thickness and having the same dip. Such a relationship between strata is said to be a conformable one.

2. *From a Partial Outcrop Cutting Three or More Contours*

In Fig. 6 the same method as that used in Fig. 5 would determine the strike and dip of the beds. However, a second method may be used here. In this instance the outcrop between limestone and shale cuts the 600 ft, 500 ft, 400 ft, 300 ft and 200 ft O.D. contours at B, C, D, E and A respectively. To determine the dip and strike of the beds, join B-A. The junction of the shale and limestone is assumed to be a plane surface, hence, there is an even gradient between B and A and between these two points the surface falls 400 ft: if the distance B-A is divided into four equal parts then the distances $B'C'$, $C'D'$, $D'E'$ and $E'A$ will each represent the horizontal distance in which the surface falls 100 ft, and C', D' and E' will have elevations of 500 ft, 400 ft and 300 ft O.D. respectively. If C is joined to C' and the line extended across the map, this will be the 500 ft strike line or stratum contour for the junction of the limestone and the shale. Other strike lines at 400 ft and 300 ft O.D. may be similarly constructed whilst lines parallel to these, and of the same spacing may be drawn

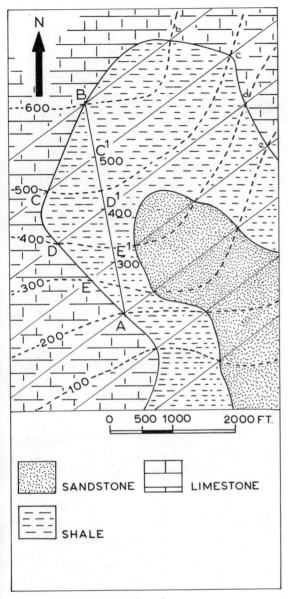

Fig. 6

across the whole map to complete the stratum contours for the surface limestone/shale.

3. *From Borehole Records*

In Fig. 7, let the boreholes be at A, B and C at topographic heights of 500 ft, 675 ft and 520 ft O.D. respectively.

Suppose a sandstone lies at the surface at A and at depths of 675 ft and 320 ft in the boreholes B and C.

The surface of the sandstone will then occur as follows: at 500 ft O.D. at A; at zero elevation at B and at 200 ft O.D. at C.

If the surface of a bed of sandstone be regarded as plane then the surface falls 300 ft from A to C; 500 ft from A to B and 200 ft from C to B.

If AC be divided into three equal parts— AA', $A'A''$ and $A''C$—each part will represent the horizontal distance in which the surface of the sandstone bed falls 100 ft in elevation in the direction of C and A, A', A'' and C will lie at 500 ft, 400 ft, 300 ft and 200 ft elevation on the surface of the sandstone.

Also if AB is divided into five equal parts— Aa, aa', $a'a''$, $a''a'''$ then a, a', a'', a''' and B will lie at 400 ft, 300 ft, 200 ft and 0 ft O.D. (sea level) on the surface of the sandstone bed.

From similar reasoning it can be seen that C' lies at an elevation of 100 ft O.D. on the surface of the sandstone bed.

A consideration of the above shows that a and A' each lie at an elevation of 400 ft O.D. on the same surface of the sandstone bed and that a line joining them would be a straight contour line or strike line at 400 ft O.D. on the surface of the sandstone: other strike lines can be drawn through $a'A''$ at 300 ft O.D.; through $a''C$ at 200 ft O.D.; through $a'''C'$ at 100 ft O.D.; lines through A and B parallel to these strike lines are the 500 ft and 0 ft (sea level) strike lines respectively.

At A, through which the 500 ft strike line passes, the sandstone surface outcrops and is overlain by a bed of shale. That the sandstone is *overlain* by the shale and not *underlain* by it may be proved by a consideration of the dip of the sandstone bed. The strike lines on its surface fall in elevation towards the southeast and this is therefore the direction of dip. The horizontal distance between each strike line is 500 ft, hence, the bed has a gradient of 1 in 5 or a dip of 11°20′ to the south-east.

Examples of the use of this method are given in the section "To Complete the Outcrops of Beds from Partial Outcrops".

TO DETERMINE THE DIP AND STRIKE OF A ROCK SERIES FROM OUTCROPS ON A GEOLOGICAL MAP

Figure 8 is a sketch of a part of Southern England where erosion has exposed junctions between the three subdivisions of the Chalk. (The sketch is taken from the Geological

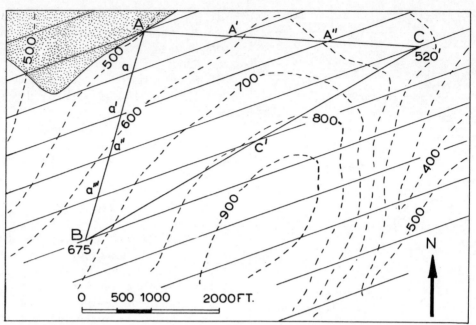

N.B. SHALE SHADING OMITTED SO AS NOT TO OBSCURE THE STRIKE LINES
FIG. 7

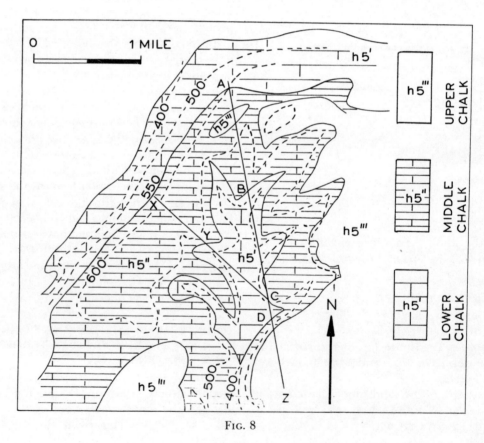

FIG. 8

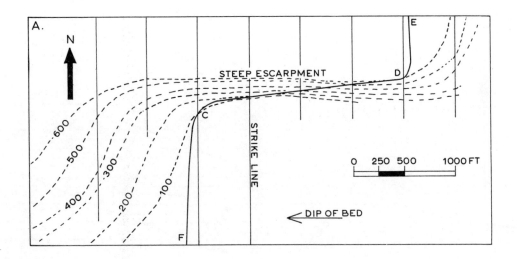
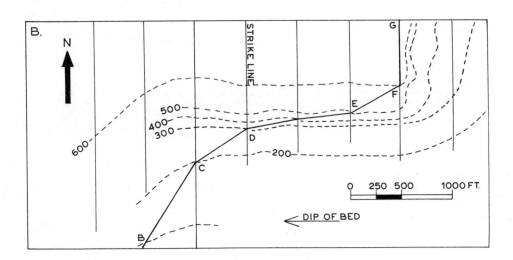
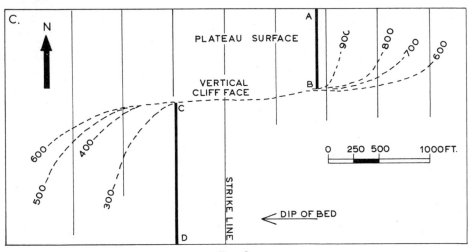

Fig. 9

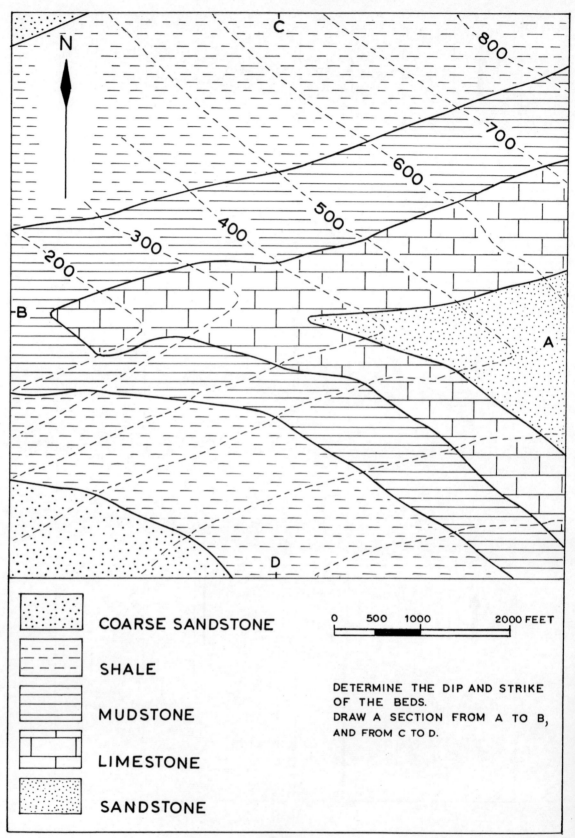

Fig. 10

Survey Map 1 inch to 1 mile, No. 282—Devizes.) The map is to such a small scale that values for the dip, obtained by the construction described below, will have no great accuracy. However, the amount of dip, in a specific direction may be obtained graphically as follows: *A*, *B* and *C* are all points on the junction plane between the Lower and Middle Chalk. If a topographic section is drawn from *A* to *Z* and, to the same vertical scale as the horizontal scale of the map, and *A*, *B* and *C* are plotted on that section and joined together, it is possible to measure the dip of the plane directly in the direction *A* to *Z*. Similarly if *X*, *Y* and *C* are plotted and joined, the dip in this direction may also be measured directly from the section. The direction and amount of true dip may then be obtained graphically or by calculation as set out in Appendix III.

OUTCROP TREND AND FORM IN RELATION TO TOPOGRAPHY

Variation in topography has a profound effect on the form of the outcrop shown on a geological map. In order to demonstrate this, three figures are discussed below in each of which the outcrop of the same bed is examined with different topographic forms.

Figure 9 shows three maps (A, B and C) in which the effect of the topography on the outcrop of a seam of coal is shown. In each map the direction of the dip of the beds is shown by an arrow. The amount of dip may be calculated by constructing the strike lines in each map. Figure 9A shows a steep escarpment with an even gradient: the even gradient is shown by equally spaced contours in the escarpment. The outcrop of the coal seam plots as a straight line in the directions *D-E* and *C-F* since the bed is outcropping on a plateau. In these positions the outcrop marks the direction of strike. The outcrop from *D* to *C* is also a straight line, traversing the escarpment from a higher to a lower topographic level in the general direction of the dip.

On a completely vertical cliff face as in Fig. 9C the outcrop between *B* and *C* would not be seen on the map and there would be an apparent gap between *B* and *C*. In this map also the outcrops plot as straight lines in the direction of strike along *A-B* and *C-D* on the plateau surfaces.

In Fig. 9B the seam outcrops as a straight line on the plateau surfaces as at *G-F*. The outcrop becomes more sinuous between *F* and *B* as the slope of the ground changes as indicated by the varying distances between the contours. Notice that as the contour lines become further apart the outcrop trends more towards the direction of strike whilst with increasing closeness the trend is in the direction of the dip. Figures 10, 11 and 12 are exercises on dip and strike determination.

PART 2

THE DETERMINATION OF THE THICKNESS OF A BED OF ROCK

In Figs. 13, 14 and 15 three frequently occurring situations are illustrated which show the relation between the dip of a bed and the slope of the ground. In each case strike lines have been inserted which trend from north to south: the dip, in each instance, is due east.

1. Consider Fig. 13 in which a bed of shale dips in the opposite direction to the slope of the land as seen from the section.

DA = the direction along the dip between the bottom and top of the bed of shale,

DE = the projection on the section of DA on the map,

AB = the apparent or vertical thickness of the shale,

BC = the true thickness of the shale,

θ = the angle of true dip of the shale.

Then the vertical thickness AB = the difference in elevation between D and A plus EB,

i.e. $$AB = AE + DE \tan \theta$$

Now since $$E\hat{D}B = C\hat{B}A = \theta$$

the true thickness is given by the formula

$$CB = AB \cos \theta$$

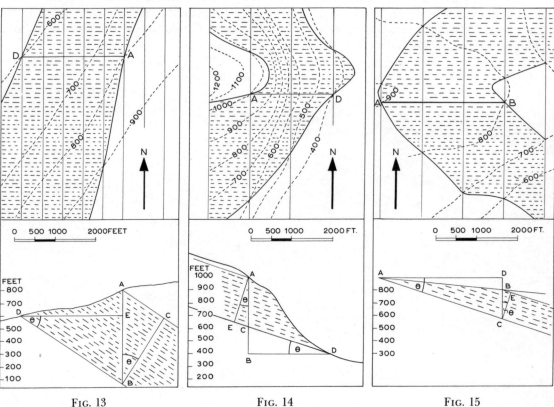

Fig. 13   Fig. 14   Fig. 15

2. Consider Fig. 14 where the bed of shale dips in the same direction as the ground surface, but at a smaller angle.

AD = the direction along the dip between the bottom and top of the bed of shale,
BD = the projection on the section of AD on the map,
AC = the apparent or vertical thickness of the shale,
AE = the true thickness of the shale,
θ = the angle of true dip of the shale.

Then the vertical thickness AC = the difference in elevation between D and A minus CB,

i.e. $\qquad AC = AB - BD \tan \theta$

Now since $\qquad B\hat{D}C = E\hat{A}C = \theta$

the true thickness is given by the formula

$$AE = AC \cos \theta$$

3. Consider Fig. 15 where the bed of shale dips in the same direction as the slope of the ground, but at a greater angle.

AB = the direction along the dip between the bottom and top of the bed of shale,
AD = the projection in the section of AB on the map,
CB = the apparent or vertical thickness of the shale,
CE = the true thickness of the shale,
θ = the angle of true dip of the shale.

Then the vertical thickness $BC = CD$ minus the difference in elevation between A and B,

i.e. $\qquad BC = DC - DB$
$\qquad\qquad = AD \tan \theta - DB$

Now since $\qquad D\hat{A}C = E\hat{C}B = \theta$

the true thickness is given by the formula

$$EC = CB \cos \theta$$

In all three cases it is seen that the thickness could be obtained by graphical means measuring directly from the section: the mathematical way is quicker and involves less construction.

Measurement has been made along the direction of true dip; should the measurement be made along a direction of apparent dip then that dip would first have to be calculated as shown in Appendix III.

PART 3

FOLDS

In previous pages the strata discussed have had a dip in one direction and are, therefore, structurally uniclinal.

Compressive forces developed within the earth's surface have caused many strata to be pushed into upfolds and downfolds of varying complexity. The simplest type of fold is shown in Fig. 16 where the bed is

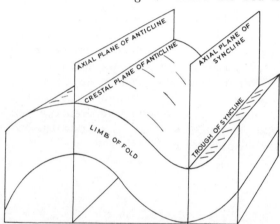

SYMMETRICAL ANTICLINE AND SYNCLINE WITH AXIAL PLANES AND CRESTAL PLANES COINCIDENT

FIG. 16

folded into an anticline (upfold) and a syncline (downfold). In this example the folds are symmetrical and hinge about a vertical plane with the limbs of the fold equally disposed on either side of it. The axial plane is the vertical plane about which the dip of the beds changes in direction and often in amount. In symmetrical folds the axial plane coincides with the crest of the fold.

TYPES OF FOLD AND THEIR RECOGNITION ON MAPS

Figure 17 is the map and section of two symmetrical anticlines and two synclines. Strike lines drawn through aa', bb', cc', dd' and ee' on the shale/sandstone junction show that the part of the junction crossed by the strike lines aa', bb' and cc' dips to the west whilst that crossed by dd' and ee' has an easterly dip: this opposed dip indicates a syncline. Since, by definition, the dip is at right angles to the strike, the "measured" horizontal distance between two of these strike lines is the horizontal equivalent for a descent of 100 ft in elevation of the bed. In the present instance the horizontal distance between the strike lines is 200 ft so that the dip is 21° 48′, or expressed as a gradient, 1 in 2·5. As all the strike lines are equally spaced over the whole map, the folds are symmetrical, as shown in the section along X-Y.

Figure 18 shows asymmetrical folds. Strike lines drawn through ff', ee' and dd' indicate a westerly dip of 1 in 1·25 when expressed as a gradient or, as an angle, 38° 44′; those drawn through gg', hh' and k' indicate an easterly dip of 1 in 2·5, as a gradient, or, as an angle, 21° 48′. The structure is, therefore, synclinal with the eastern limb of the fold being the steeper. Other asymmetrical anticlines and synclines are recognised by drawing strike lines over the whole of the map; the section along X-Y depicts the structure shown in the map.

Note: In both Figs. 17 and 18 there could be doubt as to the direction along which the strike lines should be drawn. Take as an example, Fig. 17. A strike line is a direction along which the bed maintains a constant elevation; it also joins points of like elevation on the same surface of the same bed. Points d and c, e and b and f and a all fulfil the latter part of this definition, and it may be imagined that strike lines could be constructed by joining each similar pair of points. The elevation is not maintained, however, on the

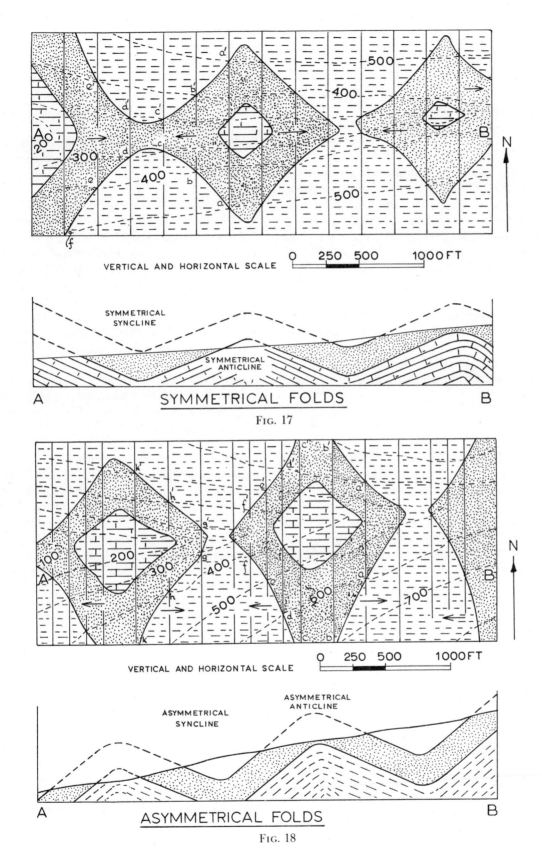

SYMMETRICAL FOLDS

Fig. 17

ASYMMETRICAL FOLDS

Fig. 18

17

same surface of the same bed along the directions *dc*, *eb* or *fa*. Further, if these directions were selected as strike lines they would not necessarily be parallel to each other, nor would they, as is usual in short distances, be equally spaced.

Overfold

In Fig. 19 along *X-Y* the beds are repeated about the arcuate outcrop of the sandstone. Strike lines drawn on the sandstone/marl boundary reveal the strike to be from north to south; the strike lines *tt'*, *ss'* and *vv'* show the dip to be westwards at a gradient of 1 in 5 or, at an angle of 14°10', whilst strike lines *lm*, *no* and *pr* indicate a westward dip of 1 in 2·5 or 21°48'. The beds lying west of the sandstone outcrop maintain the latter dip whilst those to the east keep the former dip. This constancy in direction but variation in amount of dip is characteristic of overfolded beds. This structure is shown in the section along *X-Y*.

Recumbent Fold

In Fig. 20 a consideration of the junction coarse sandstone/limestone in the central and south-eastern part of the map show it occurring at two elevations, one at 1050 ft and the other at 850 ft, each follows the position of the contour of that value: this indicates that both junctions are horizontal so that at 1050 ft the limestone appears above the coarse sandstone whilst, at 850 ft, it appears below it. Strike lines such as *ab* and *cd* in this same junction, in the western part of the map indicate a dip to the west of 1 in 2·5 or 21°48' on the 1050 ft junction; the continuation of this dip would bring the 1050 ft junction to intersect the 850 ft junction. The interpretation of this disposition is shown in the section as a typical recumbent fold such as the spectacular one seen on the face of the Matterhorn.

Isoclinal Folds

Figure 21 shows the map and section of a series of isoclinal folds. Strike lines, such as *hh'*, *gg'*, *ff'*, etc., drawn across the map indicate that the beds have a uniform westerly dip. If each similar type of bed, for example the sandstone, can be shown to be of the same age, then the interpretation would be as in the section along *X-Y* with a series of isoclinal folds, the axial planes of which lie parallel to each other: note that the crest of each fold is not coincident with the trace of the axial plane in this type of folding. Folds of this type are developed in the Southern Uplands of Scotland.

Plunging Folds

Often in a folded series the beds are not only folded about an axial plane, but the axis itself it tilted. Such folds are said to plunge or pitch: Figures 23 and 24 show a plunging anticline and a plunging syncline.

Figure 22 shows a series of folds plunging to the south. The diagram shows the dip section appearing as in any other folded series, with no indication of the plunging nature of the folds. The north to south section along the plunge shows the beds dipping with the plunge and not horizontal as in a strike section in non-plunging folds.

Figure 25 is a map and section of a group of plunging folds. To indicate clearly the nature of these folds the strike lines or stratum contours have been drawn. Characteristically the outcrops of the plunging series lie en echelon as shown in the map, the outcrops opening out in the direction of plunge in the synclines and closing with the plunge in the anticlines. In this, and comparable situations, strike lines will no longer remain parallel across large areas, but will converge in relation to the closure of folds. This is an important consideration in the construction of strike lines on this map and subsequent similar maps.

Consider the junction of the shale and of the mudstone along *AB* and along *CD*. Strike lines are drawn on the *AB* junction by joining *bb'*, *aa'* and *ww'* and these show that the beds dip in a south-easterly direction with a gradient of 1 in 4. Strike lines drawn on the *CD* junction through *cc'*, *ee'*, *ff'* and *k* show this junction to have a south-westerly dip with a gradient of 1 in 2. The strike lines *aa'* and *cc'*, which are both at 400 ft O.D. cross each other to the north and so form a chevron-shaped strike

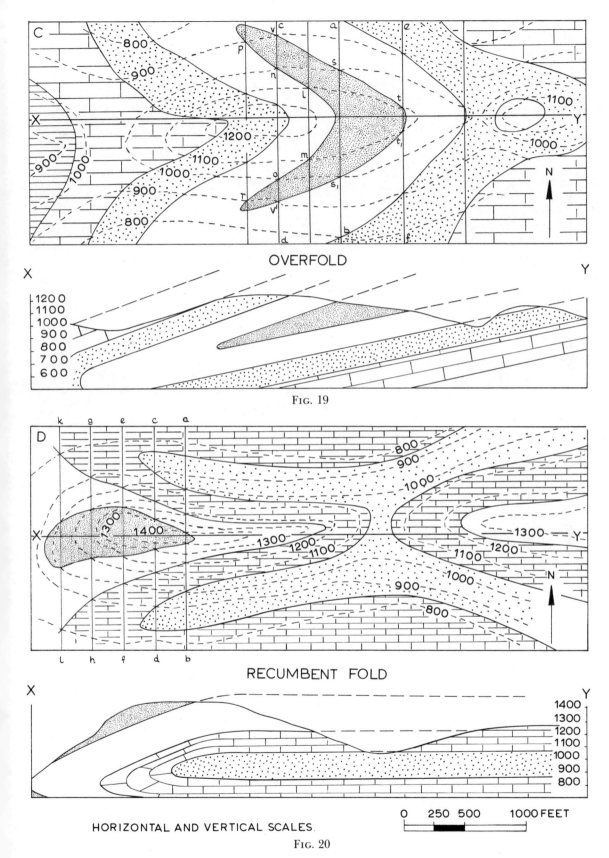

OVERFOLD

Fig. 19

RECUMBENT FOLD

HORIZONTAL AND VERTICAL SCALES.

Fig. 20

19

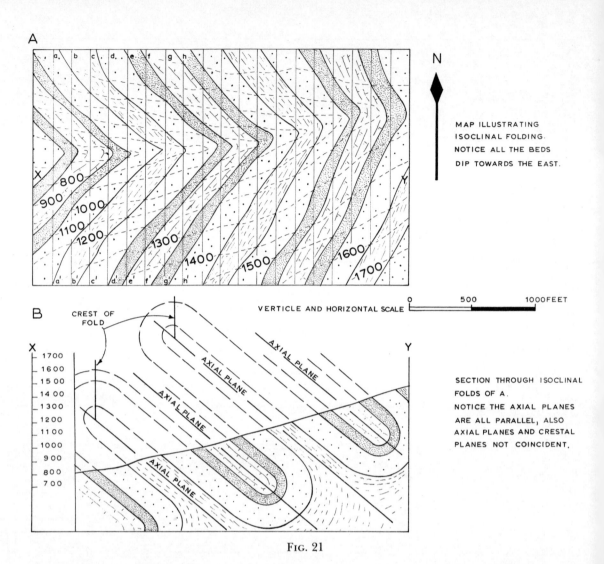

Fig. 21

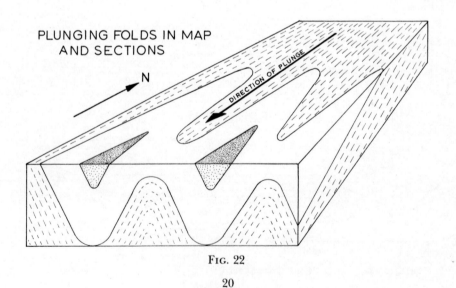

Fig. 22

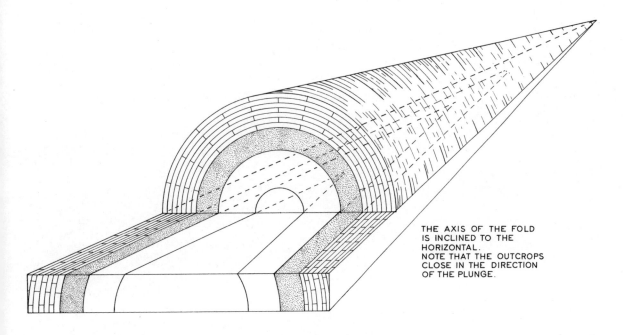

PLUNGING ANTICLINE
FIG. 23

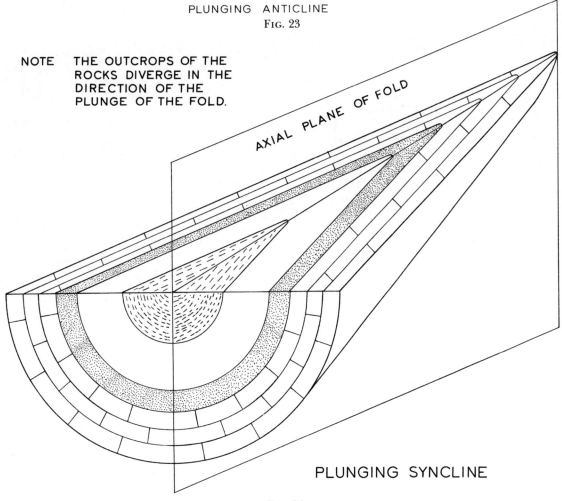

PLUNGING SYNCLINE

FIG. 24

line: other strike lines run parallel to this through *b, d, g, h* and *k, f* and *e* to produce an en echelon pattern in the syncline.

These two boundaries, therefore, dip towards each other whilst the axis between and *nn'* in the former and through *tt', pp', ss'* and *r* in the latter and that they also form an en echelon pattern, but with the dips away from the northerly trending axis of the fold; this is an *asymmetrical plunging anticline*.

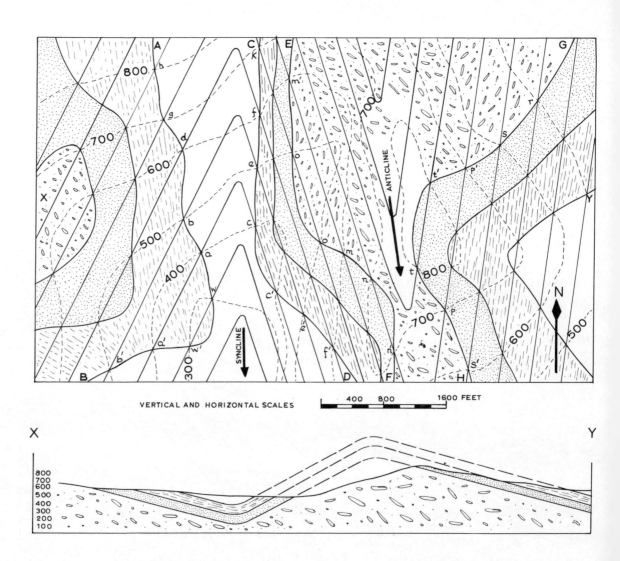

MAP AND SECTION OF A SERIES OF PLUNGING FOLDS
Fig. 25

them dips southwards. This structure is an *asymmetrical plunging syncline*.

Further consideration of the conglomerate/sandstone junctions *EF* and *GH* shows that strike lines can be drawn through *oo', mm'*

Notice that the section, drawn across the folds, does not give any indication of their plunging character.

Figures 26, 27, 28 are exercises to be attempted by the student.

22

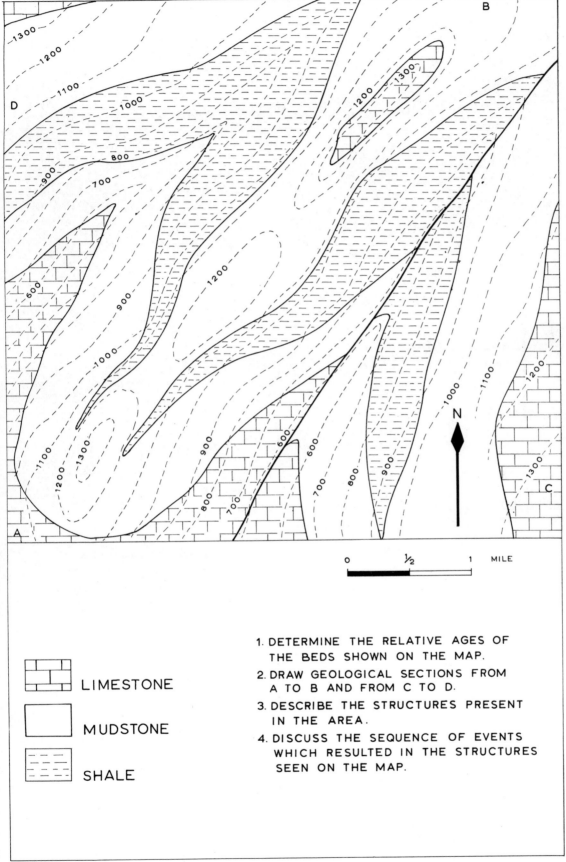

Fig. 26

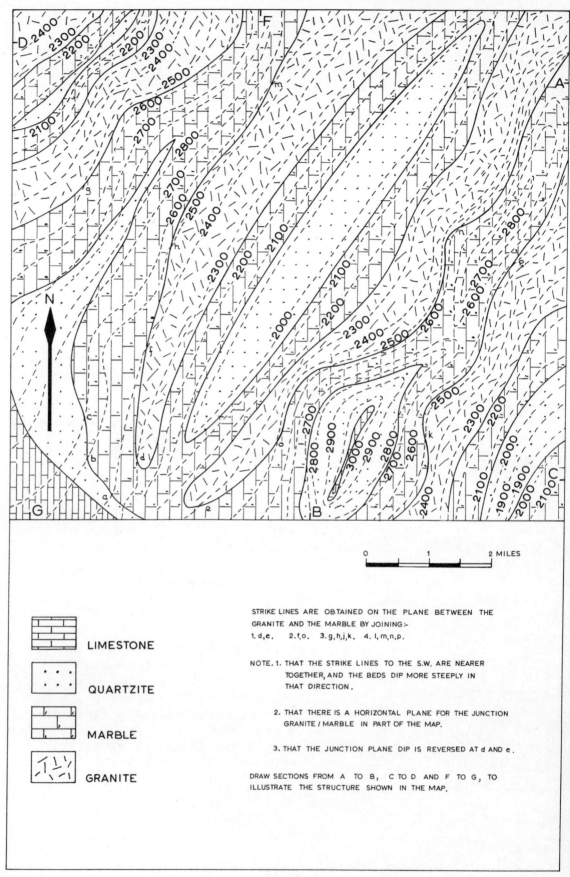

FIG. 27

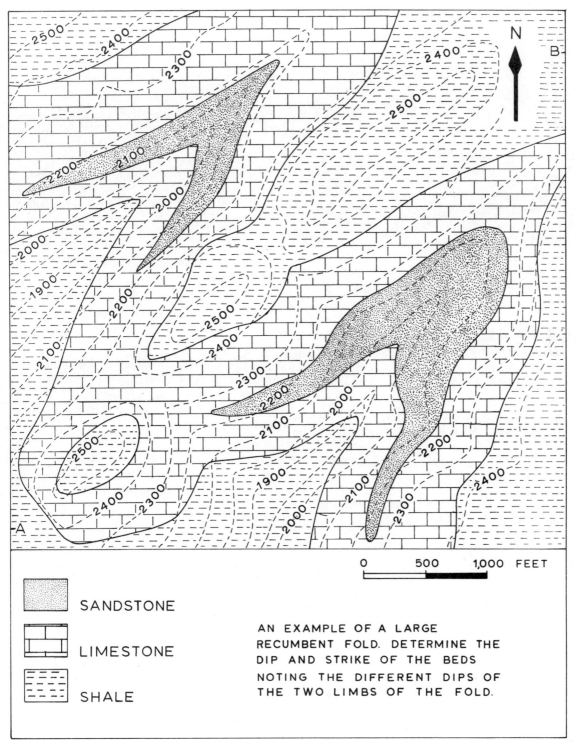

FIG. 28

PART 4

FAULTS

Faults are breaks in rock masses producing an observable displacement on either side of the surface of fracture. The surface along which the movement takes place is referred to as the fault plane; such surfaces are, however, usually curved and irregular and the movement affects a zone rather than a single surface. Tensional, compressional and torsional forces operate in the formation of faults and much may be learned of the nature of the forces responsible for the breaks from the patterns made by faults and their associated fractures. No attempt will be made to discuss the wider problems of the origin of faults and their interpretation in this book. In the following pages will be set out:

A. The descriptive terminology of faults.
B. The broad classification of faults.
C. The effect of faults on the outcrops of beds.
D. Practical examples of the effect of faulting.
E. Methods for the determination of the throw of a fault.

A. DESCRIPTIVE TERMINOLOGY OF FAULTS (Fig. 29)

(i) THE FAULT PLANE. The surface along which a series of rocks fracture is the fault plane: in Fig. 29 *MNPO* is the fault plane.

(ii) THE HANGING WALL AND THE FOOT WALL. The fault plane is usually inclined and that part of the rock mass lying *above* the fault plane is the hanging wall and that *below* is the foot wall.

(iii) THE UPTHROW AND DOWNTHROW SIDES OF THE FAULT. These terms refer to the relative movement of the rock masses on either side of the fault plane; the side where

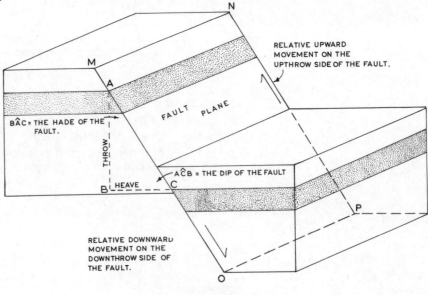

FIG. 29

the movement has apparently been downwards is the *downthrow side* and that in which the movement has apparently been upwards is the *upthrow side*. The downthrow side of a fault is commonly indicated on a map by a short line at right angles to the trace of the fault plane.

(iv) THE DIP OF A FAULT. This is the angle between the fault plane and the horizontal and is shown in Fig. 29 by the angle $A\hat{C}B$.

(v) THE HADE OF A FAULT. This is the angle between the fault plane and the vertical: it is the complement of the dip and in Fig. 29 is the angle $B\hat{A}C$.

(vi) THE THROW OF THE FAULT. The vertical displacement of the severed ends of a faulted bed is the *vertical throw* of the fault (Fig. 29, AB). This displacement must be distinguished from the *stratigraphical throw* which is the amount of separation of a faulted bed normal to the bedding plane (Fig. 30, X-Z). The two values are related by the formula

$$S = V \cos \theta,$$

where S is the stratigraphical separation, V the vertical separation and θ the true dip of the beds.

considerable importance in mining where the separation of this kind means loss of the seam or vein.

(viii) DIRECTION OF SLIP OR MOVEMENT. When two blocks of rock are broken by a fault the movement may be wholly in a horizontal manner parallel to the fault plane. This is the *strike slip*; it may be wholly along the direction of maximum dip of the fault plane, this is *dip slip*, or a combination of the two when the movement is the *oblique slip*: Fig. 31 illustrates these directions of movement.

B. THE BROAD CLASSIFICATION OF FAULTS

The problem of classification is complex and should have its basis in the study of fault genesis. Here, however, the classification is based on characteristics which are relatively easily determined and are fundamentally descriptive. The bases for this classification are:

1. *The Apparent Relative Movement of the Hanging Wall and the Foot Wall*

Figure 32A shows a faulted block in which the hanging wall has had an apparent downward movement; the result is that the fault

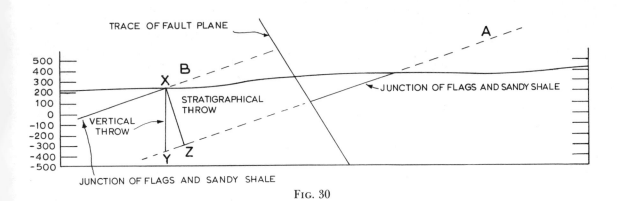

FIG. 30

(vii) THE HEAVE OF A FAULT. The horizontal displacement of the severed ends of a bed of rock is called the *heave* of the fault: this is illustrated by BC in Fig. 29. The heave is of

plane hades to the downthrow side; such faults are described as *normal faults*.

Figure 32B shows a faulted block in which the foot wall has had an apparent downward

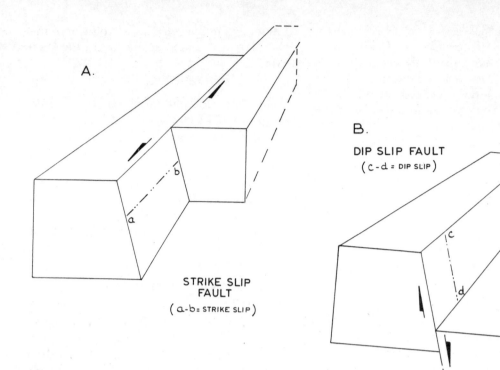

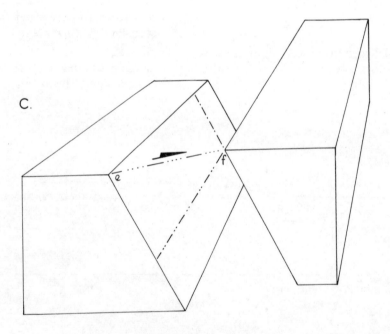

OBLIQUE SLIP FAULT (e-f = OBLIQUE SLIP)
WHICH IS THE RESULTANT OF THE
DIP AND STRIKE SLIP.

Fig. 31

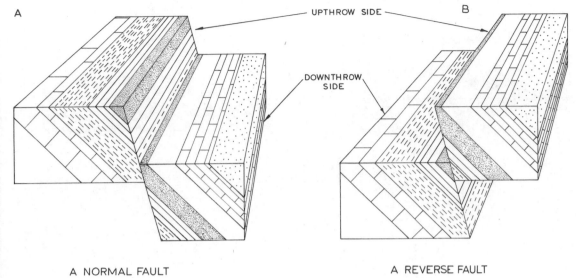

A NORMAL FAULT   A REVERSE FAULT

Fig. 32

movement. The fault plane, therefore, hades towards the upthrow side: this is a *reverse fault*.

2. *The relation of the Dip and Strike of the Fault Plane to that of the Beds affected*

Figure 33 shows three types of fault founded on this criterion. A shows the strike of the fault parallel to the dip: this is a *dip fault*. B shows the strike of the fault plane parallel to that of the strike of the beds: this is a *strike fault*. C shows the strike of the fault plane oblique to the strike and dip of the beds; this is called an *oblique fault*.

A little experience will teach that a fault seldom is so clear cut in its relation to the beds it fractures as is indicated above, but there will usually be a greater movement in one direction than in another and a fault may be named in accordance with the direction in which the displacement is greatest.

3. *The Direction of Movement on the Fault Plane*

Figures 34A, B and C show the principle directions in which the relative movement on a fault plane may take place. In A the movement coincides with the dip of the fault and so is described as *dip slip*. B shows the movement along the strike; this is *strike slip*, whilst in C the movement is a combination of dip slip and strike slip and so is described as *oblique slip*.

4. *Types of Movement*

The movement in the faults so far named has

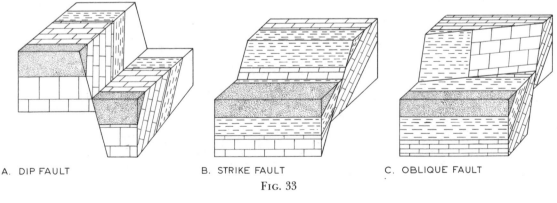

A. DIP FAULT   B. STRIKE FAULT   C. OBLIQUE FAULT

Fig. 33

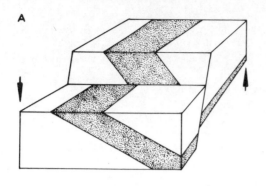
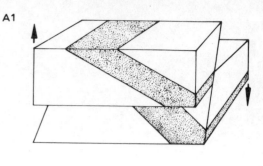
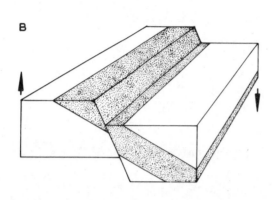
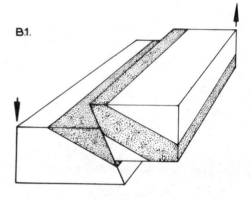
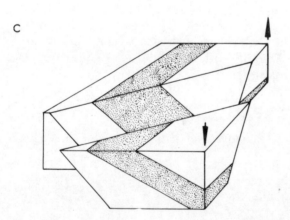
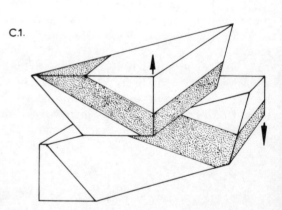

A. NORMAL DIP FAULT   B. NORMAL STRIKE FAULT.   C. NORMAL OBLIQUE FAULT.
A.1. REVERSE DIP FAULT   B.1. REVERSE STRIKE FAULT   C.1. REVERSE OBLIQUE FAULT

NOTE. IN THE NORMAL FAULT THE HADE OF THE FAULT PLANE IS TOWARDS THE DOWNTHROW SIDE.
IN THE REVERSE FAULT THE HADE OF THE FAULT PLANE IS TOWARDS THE UPTHROW SIDE.
THE ARROWS INDICATE THE RELATIVE DIRECTION OF THE MOVEMENT.

FIG. 34

A

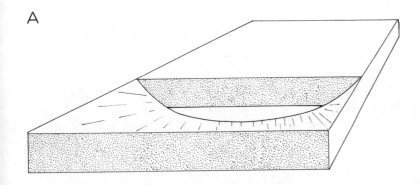

A SAG FAULT. SAG FAULTS ARE ASSOCIATED WITH AREAS OF GENTLE TECTONIC DISTURBANCE AND WITH AREAS OF MINERALISATION. NOTICE THAT THERE IS A VARYING THROW OF THE FAULT.

B

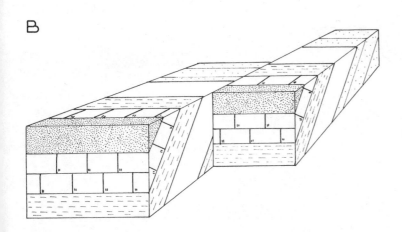

A TEAR FAULT. THE BEDS HAVE MOVED ONLY IN A HORIZONTAL DIRECTION. IN REALITY FEW FAULTS HAVE ONLY HORIZONTAL MOVEMENT.

C

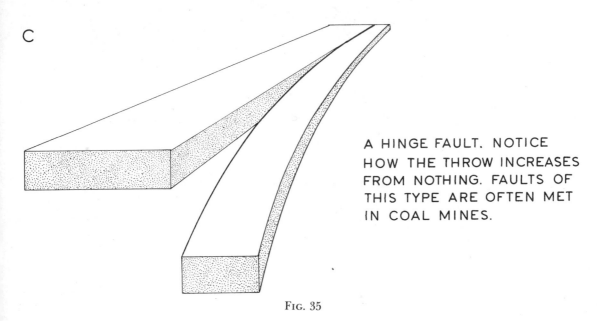

A HINGE FAULT. NOTICE HOW THE THROW INCREASES FROM NOTHING. FAULTS OF THIS TYPE ARE OFTEN MET IN COAL MINES.

FIG. 35

been in equal amounts over a wide planar surface. Three further types have such movement; they are: (i) The tear fault, where the movement is almost wholly horizontal—Fig. 35B. Such faults are common in South Wales—see the Ammanford Sheet 230, Geological Survey Map 1 inch to 1 mile. (ii) The horst, where as a result of two or more near parallel faults a block of country is upthrust—Fig. 36A. This kind of structure is common in the Vosges. (iii) The graben or rift valley where a block of country founders between groups of parallel faults—Fig. 36B. The African Rift Valley, having a total length of 1800 miles and an average

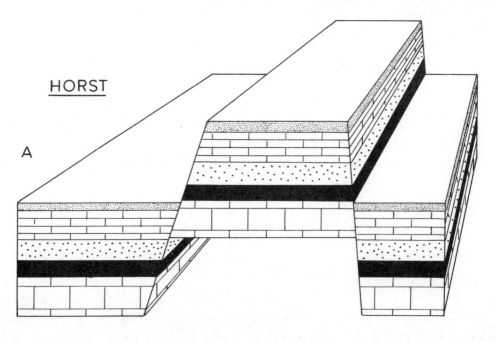

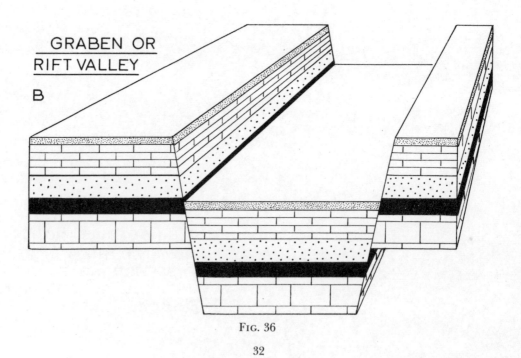

Fig. 36

width of 20 miles, is the classic example of this, but the Midland Valley of Scotland and the Vale of Clwyd in North Wales are excellent examples on a smaller scale.

A further group of faults shows differential throw along the fault plane. Figure 35A shows a sag fault where the throw increases from the undisturbed margins to the maximum central sag.

Figure 35C shows a hinge fault where the throw increases from the plane of the hinge. Figure 37 is a swivel fault where sliding has taken place on a low angle fault plane. This is in effect a rotational effect which can be seen in the Carboniferous Limestone of the Gower Peninsula in South Wales and is associated with many major thrust areas.

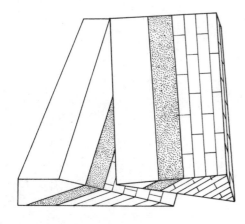

SWIVEL FAULT
Fig. 37

C. THE EFFECT OF FAULTS ON THE OUTCROPS OF BEDS

Figures 38–44 show in diagram form the main effects of faulting both on simple dipping series and on folded series. A careful study of these will reveal the major effects and the information so gained should be used in examining faulting in the 1 inch to 1 mile Geological Survey Maps.

D. TWO PRACTICAL EXAMPLES OF THE EFFECT OF FAULTING

Faulting is very important in everyday activities, particularly mining. Two examples are now given:

In Fig. 45A, a normal strike fault has broken a seam of coal: the severed edges of the seam have been drawn away from each other as a result of the faulting. This results in there being a zone in which coal is absent — the barren ground; this is often a region of much fragmented rock and causes difficulty in driving headings and roadways. In Fig. 45B the seam is broken by a reverse strike fault: here the severed ends of the seam have ridden over each other so that a borehole put down from A would pass through the same seam twice. Both these types of movement are well illustrated in the South Wales Coalfield.

E. THE DETERMINATION OF THE VERTICAL THROW OF A DIP FAULT

In Fig. 46 the bed of sandstone is broken by a dip fault *RS* which trends from north to south.

In order to find the throw of this fault, first determine the dip and strike of the sandstone by drawing strike lines on the *same* surface of the *same* bed. A line through *dd'* gives the *direction* of strike which here is east to west. Other strike lines, parallel to *dd'*, through *a*, *A*, *C*, *X* and *e* enable the direction and amount of dip to be determined. The distance between the strike lines is 200 ft and the vertical interval between them is 100 ft: the dip is therefore as a gradient, 1 in 2 to the south.

Two methods are considered below for determining the vertical throw of a dip fault.

1. *The Equal Altitude Method*

(a) Select two points, which are at the same elevation, on the same surface of the bed and on opposite sides of the fault: let these points be *A* and *C*, each at an elevation of 700 ft O.D. Join *AC*.

(b) Through *C* draw a strike line *CD* to intersect a line drawn, in the direction of dip, from *A* at *B*.

(c) Use the scale on the map and measure *AB* which is 400 ft. Since the dip is 1 in 2 to the south, the bed will fall 200 ft in elevation in 400 ft horizontal distance. *A* is at 700 ft O.D. so that a fall of 200 ft to *B* means that *B* will lie at 500 ft O.D., but *B* and *C* lie on the same strike direction so that in order for *C* to be at

A   PRE FAULTING

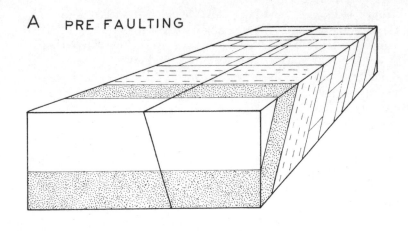

A DIP FAULT i.e. PARALLEL WITH THE DIRECTION OF THE DIP OF THE BEDS.

B   POST FAULTING

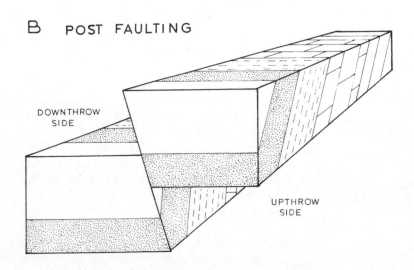

THE FAULT IS A REVERSE FAULT i.e. THE HADE OF THE FAULT IS TOWARDS THE UPTHROW SIDE.

C   POST EROSION

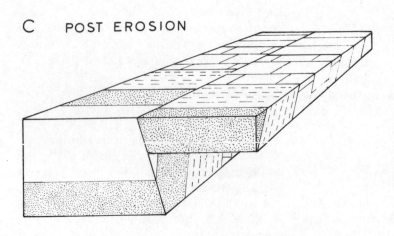

THE PLAN OR MAP OF THE FAULTED AREA. THE BEDS HAVE APPARENTLY MOVED IN THE DIRECTION OF DIP ON THE UPTHROW SIDE.

REVERSE DIP FAULT.
Fig. 38

A. PRE FAULTING

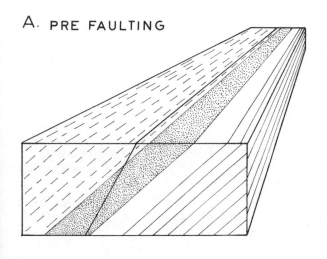

A STRIKE FAULT HADING WITH THE DIP OF THE BEDS.

B. POST FAULTING

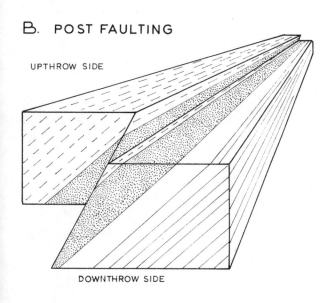

THE FAULT IS A REVERSE STRIKE FAULT i.e. THE HADE OF THE FAULT IS TOWARDS THE UPTHROW SIDE.

C. POST EROSION

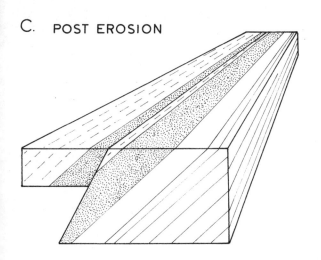

THE PLAN OR MAP OF THE FAULTED AREA. NOTICE THAT AS A RESULT OF THE FAULTING BEDS HAVE BEEN REPEATED AT THE SURFACE.

A REVERSE STRIKE FAULT HADING WITH THE DIP OF THE BEDS RESULTS IN THE REPETITION OF BEDS.

FIG. 39

A. PRE FAULTING

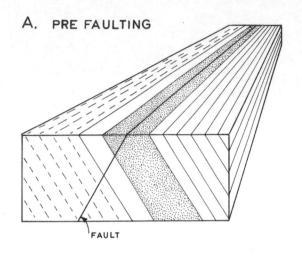

A STRIKE FAULT—THIS IS A FAULT THE TRACE OF WHICH IS PARALLEL OR NEARLY PARALLEL TO THE STRIKE OF THE BEDS. THE FAULT IS HADING AGAINST THE DIP OF THE BEDS.

B. POST FAULTING

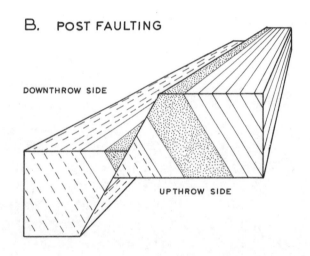

THE FAULT HADES TOWARDS THE DOWNTHROW SIDE AND IS THEREFORE CALLED A NORMAL FAULT.

C. POST EROSION

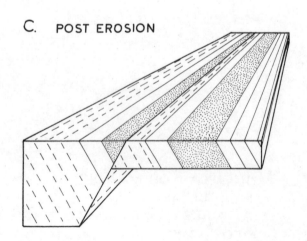

THE PLAN OR MAP OF THE FAULTED AREA. NOTICE THAT AS A RESULT OF THE FAULTING THE BEDS HAVE BEEN REPEATED AT THE SURFACE.

A NORMAL STRIKE FAULT HADING AGAINST THE DIP OF THE BEDS RESULTS IN THE REPETITION OF THE BEDS.

FIG. 40

A. PRE FAULTING

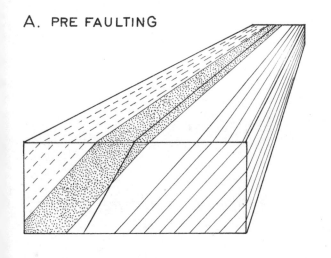

A STRIKE FAULT HADING WITH THE DIP OF THE BEDS.

B. POST FAULTING

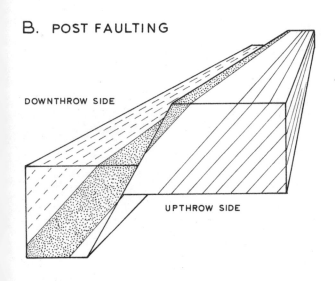

THE FAULT IS A NORMAL STRIKE FAULT, i.e. THE HADE OF THE FAULT IS TOWARDS THE DOWNTHROW SIDE.

C. POST EROSION

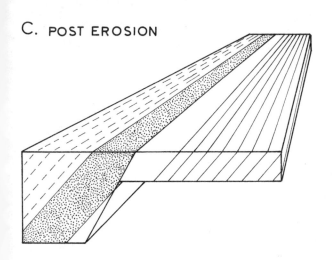

THE PLAN OR MAP OF THE FAULTED AREA. NOTICE THAT AS A RESULT OF THE FAULTING PART OF THE SANDSTONE BED IS CUT OUT.

A NORMAL STRIKE FAULT HADING WITH THE DIP OF THE BEDS RESULTS IN THE CUTTING OUT OF PART OF A BED OR ENTIRE BEDS.

FIG. 41

A. PRE FAULTING

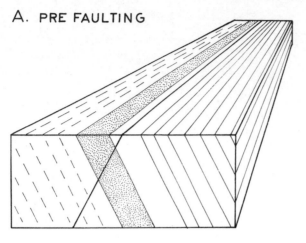

A STRIKE FAULT HADING AGAINST THE DIP OF THE BEDS.

B. POST FAULTING

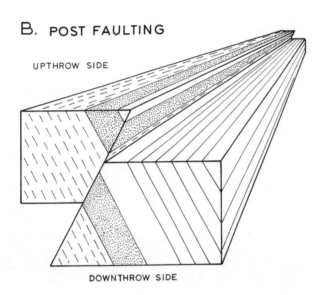

THE FAULT IS A REVERSE FAULT i.e. THE HADE OF THE FAULT IS TOWARDS THE UPTHROW SIDE.

C. POST EROSION

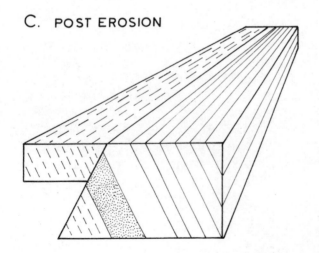

THE PLAN OR MAP OF THE FAULTED AREA. NOTICE THAT AS A RESULT OF THE FAULTING THE SANDSTONE BED HAS BEEN CUT OUT.

A REVERSE STRIKE FAULT HADING AGAINST THE DIP OF THE BEDS RESULTS IN THE CUTTING OUT OF PART OF A BED OR ENTIRE BEDS.

FIG. 42

A PRE FAULTING

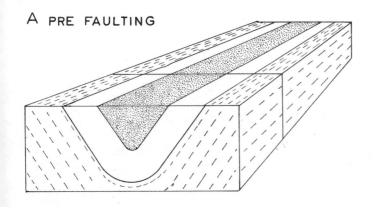

A SYMMETRICAL SYNCLINE

B POST FAULTING

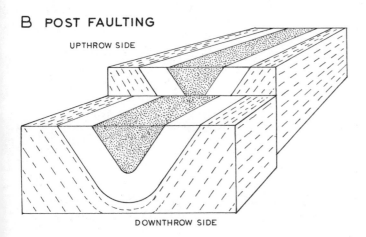

A SYMMETRICAL SYNCLINE BROKEN BY A DIP FAULT.

C POST EROSION

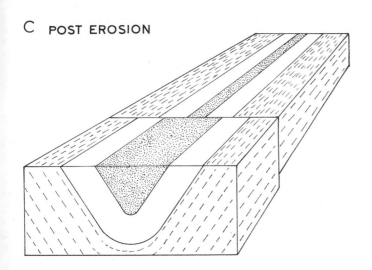

PLAN OR MAP OF FAULTED SYNCLINE – NOTICE THAT THE OUTCROPS APPEAR TO MOVE TOWARDS EACH OTHER ON THE UPTHROW SIDE.

A FAULTED SYMMETRICAL SYNCLINE.

Fig. 43

A PRE FAULTING

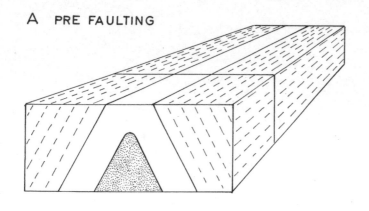

A DIP FAULTED SYMMETRICAL ANTICLINE.

B POST FAULTING
UPTHROW SIDE

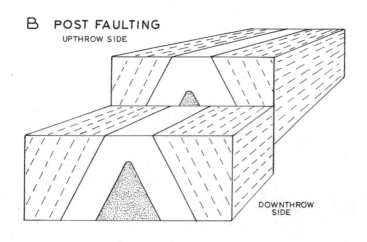

DOWNTHROW SIDE

THE DIP FAULT UPTHROWS THE ANTICLINE TO THE RIGHT.

C POST EROSION

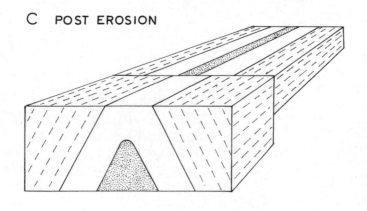

MAP OR PLAN OF THE FAULTED ANTICLINE. ON THE UPTHROW SIDE THE BEDS APPEAR TO MOVE AWAY FROM THE AXIS OF THE ANTICLINE WHEN THE UPTHROWN BLOCK IS ERODED.

SHOWING THE EFFECT OF A DIP FAULT
ON A SYMMETRICAL ANTICLINE

FIG. 44

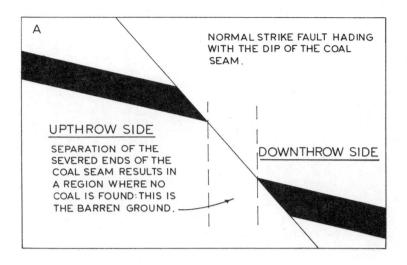

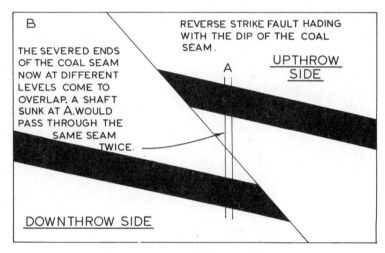

Fig. 45

700 ft O.D. there must have been an upthrow of 200 ft to the east, or producing the same effect, a downthrow of 200 ft to the west.

This result is seen in the strike section where sections along MN and OP are superimposed on each other. A is the outcrop of the top of the sandstone west of the fault and C is its outcrop to the east of the fault. Their relative elevations at 700 ft and 500 ft O.D. shows there to be a western downthrow of CC' which is 200 ft.

Most of the above measurements can be readily seen from the map in the present instance, but where there are odd distances as could occur in the direction AB or where the throw of the fault is not an even number of feet, a simple mathematical treatment is helpful and is expressed as follows:

The throw of the fault = the displacement of the bed surface along the dip × tangent of the angle of dip.

In the present case

$$CC' = 400 \times 0.5$$
$$= 200 \text{ ft}$$

2. *The Strike Method*

Consider the strike line WV which intersects the outcrops of the top of the sandstone west of the fault at XX′ and east of the fault at Y. and X′ each lie at 400 ft O.D. and Y lies at 600 ft O.D. This indicates a difference in

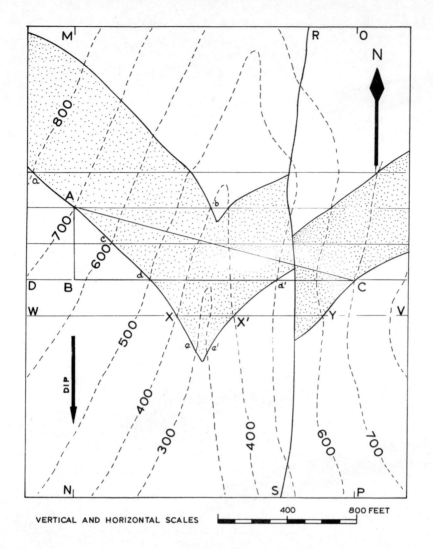
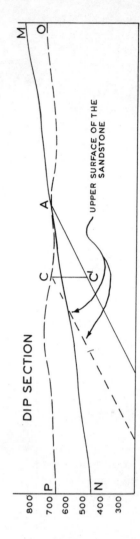
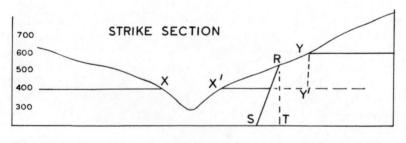

FIG. 46

THE DETERMINATION
OF THE THROW OF
A DIP FAULT.

elevation of the sandstone of 200 ft as between the two sides of the fault.

The strike section taken along *WV* illustrates this method. The beds in a strike section are horizontal so that to the west of the normal dip fault *RS* the top of the sandstone bed lies at 400 ft O.D. along the whole section: east of the fault the same surface outcrops at *Y* and the bed lies at 600 ft O.D. along the whole of this section. The vertical throw is, therefore, equal to the difference in elevation in the top of the sandstone on each side of the fault; this is Y-Y' which is 200 ft.

3. The Calculation of the Throw of a Strike Fault

Figure 47 shows a bed of sandstone broken by a reverse strike fault.

In order to calculate the effect of the fault first determine the dip and strike of the two sandstone outcrops east and west of the fault.

A line joining dd' gives the direction of strike of the eastern outcrop whilst parallel lines through a, b, c and e, on the upper surface of the sandstone, enable the dip to be calculated: since the strike lines represent vertical intervals of 100 ft and are 200 ft apart, the dip, as a gradient is 1 in 2 to the south.

A line joining gg' and with lines parallel to this through f and h, shows the dip and strike of the western outcrop to be the same as that in the eastern.

It is, therefore, apparent that the dip and strike of the sandstone bed are the same both east and west of the fault.

Consider the outcrop of the lower surface of the sandstone at A and A'. If the sandstone continued to dip from A' westwards without the intervention of the fault it would have fallen to an elevation of 300 ft O.D. at A as is seen from a consideration of the strike lines. However, the outcrop at A is at 800 ft O.D. so that the fault has upthrown the sandstone 500 ft on the western side of the fault.

The throw may also be calculated by the equal altitude method as follows:

(a) Select two points of like elevation on the same surface of the bed to east and west of the fault. Let the two points be d' and g'. Join d'-g'.

(b) Measure the horizontal distance d'-g' which equals 1131 ft.

(c) Draw $d'd''$ in the direction of dip to intersect the strike line through g' at d''.

(d) The direction $d'g'$ is a direction of apparent dip: measure the angle between this and the direction of true dip: this angle is 27°.

Now the throw of the fault may be calculated as follows:

The throw of the fault between d' and g' = $d'd'' \times$ tan of true dip along $d'd''$ = 1 in 2 = 1000 ft $\times$ 0·5 = 500 ft.

The throw of the fault is, therefore, 500 ft upthrow to the west of the fault.

Or throw from d' to $g' = d'g' \times$ tan of apparent dip along $d'g'$.

Now tan of apparent dip along $d'g'$ = tan of true dip $\times$ cosine of the angle between the true and apparent dip

$$= 0·5 \times \tan 27°$$
$$= 0·5 \times 0·8910$$
$$= 0·445$$

Now throw along $d'g' = d'g' \times$ tan of apparent dip on $d'g'$

$$= 1131 \text{ ft} \times 0·445$$
$$= 503 \text{ ft}$$

This gives the same answer, with a small allowable error, of 503 ft upthrow to the south.

4. Some Examples of Faulting to be Studied

Figure 48 depicts a normal strike fault hading against the dip of the beds.

The nature of the fault is established by drawing strike lines on the fault outcrop—lines through 2·2 and 3·3 give the direction of strike: the fault, therefore, dips at a gradient of 1 in 1 in a direction west of south.

The throw of the fault is then obtained by measuring A-B across the fault in the direction of dip: this distance is 2100 ft. Now if the junction sandy shales/flags had dipped southwards, without the intervention of the fault, it would have fallen 700 ft in the horizontal distance of 2100 ft and so would lie at -300 ft at B: but south of the fault the 300 ft strike line for this junction passes through B. Hence the fault has a downthrow of 600 ft to the north or, what is the same thing, an upthrow of 600 ft to the south of the fault.

In Fig. 49 strike lines are already drawn on the fault. The student should determine the type of fault and the throw.

Figures 50 and 51 provide exercises in the study of faults.

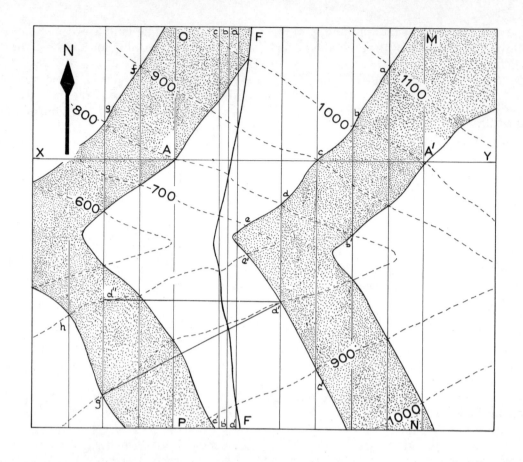

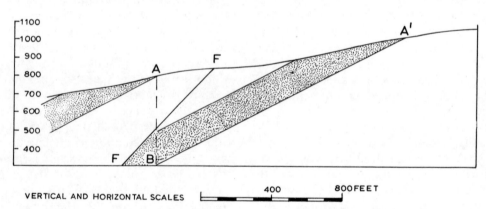

DETERMINATION OF THE THROW OF A STRIKE FAULT

Fig. 47

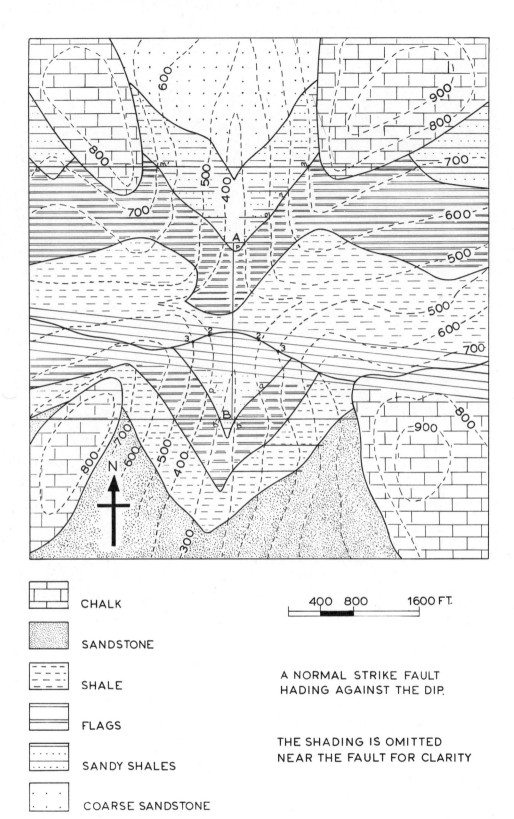

	CHALK
	SANDSTONE
	SHALE
	FLAGS
	SANDY SHALES
	COARSE SANDSTONE

A NORMAL STRIKE FAULT HADING AGAINST THE DIP.

THE SHADING IS OMITTED NEAR THE FAULT FOR CLARITY

Fig. 48

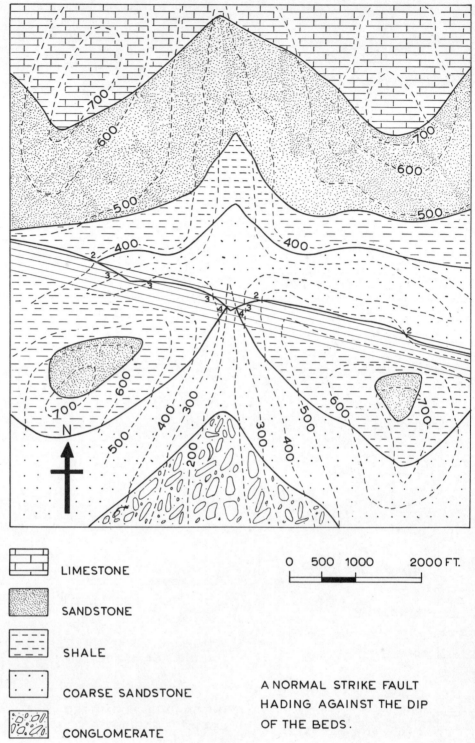

A NORMAL STRIKE FAULT HADING AGAINST THE DIP OF THE BEDS.

Fig. 49

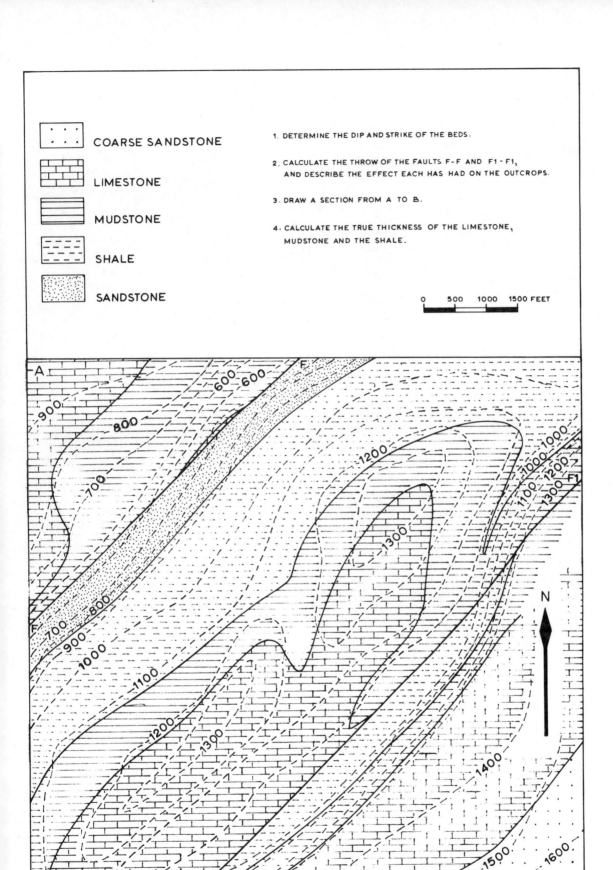

- COARSE SANDSTONE
- LIMESTONE
- MUDSTONE
- SHALE
- SANDSTONE

1. DETERMINE THE DIP AND STRIKE OF THE BEDS.
2. CALCULATE THE THROW OF THE FAULTS F-F AND F1-F1, AND DESCRIBE THE EFFECT EACH HAS HAD ON THE OUTCROPS.
3. DRAW A SECTION FROM A TO B.
4. CALCULATE THE TRUE THICKNESS OF THE LIMESTONE, MUDSTONE AND THE SHALE.

0   500   1000   1500 FEET

FIG. 50

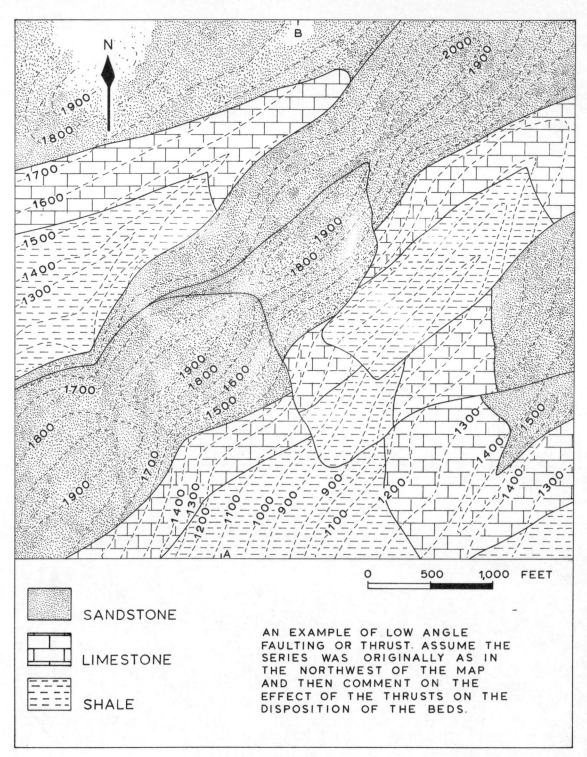

FIG. 51

PART 5

UNCONFORMITIES

In the geological succession of many areas, two or more rock groups occur, each with its own dip and, commonly, its own direction of strike. Generally each series of rocks is separated from the other by a plane of erosion or of non-deposition. Frequently, there lies at the base of the younger series a conglomerate of varying thickness and patchy in its occurrence. Some degree of discordance is usually apparent in all unconformities.

Figure 52 shows four common types of unconformity.

Figure 52A—a discordance of dip occurs between two series of rocks: in this two-dimensional figure, the strike of each series appears parallel to that of the other; the more usual situation is where amount of dip and direction of strike differ in each of the associated series. Sheet 281—Frome, shows this type of unconformity in Vallis Vale, where the Jurassic (Lias) comes to lie on the upturned edges of Carboniferous Limestone.

In Fig. 52B horizontal beds of sediment lie on the eroded surface of a granite. Sheet 156–Leicester, shows the Triassic beds lying on the eroded Mount Sorrel granite, which is probably Devonian in age.

The unconformity in Fig. 52C occurs between two series in which there is marked difference in the degree of folding. The Geological Survey of Scotland, Sheet 15—Sanquhar, illustrates this type, where north-west of the Southern Uplands Fault, the Lower and Upper Old Red Sandstone display different intensities in their folding.

In the fourth example, Fig. 52D, igneous dykes terminate against an overlying series and this, taken in conjunction with the variation in dip is a manifestation of an unconformity. The

A THE UNCONFORMITY IS SHOWN BY A DISCORDANCE OF DIP AND STRIKE BETWEEN THE TWO SERIES.

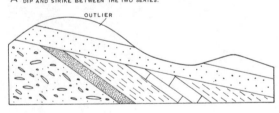

B THE HORIZONTAL BEDS ARE LYING UNCONFORMABLY ON THE ERODED SURFACE OF THE OLDER ROCK.

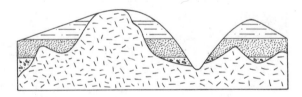

C THE UNCONFORMITY IS SHOWN BY THE UPPER SERIES EXHIBITING A LESS INTENSE FOLDING THAN THE LOWER SERIES.

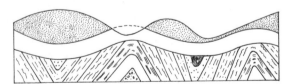

D THE UNCONFORMITY IS SHOWN BY DISCORDANCE OF DIP BETWEEN THE TWO SERIES AND BY THE RESTRICTION OF THE INTRUSIONS TO THE LOWER SERIES.

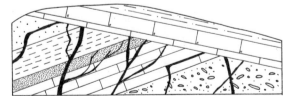

Fig. 52

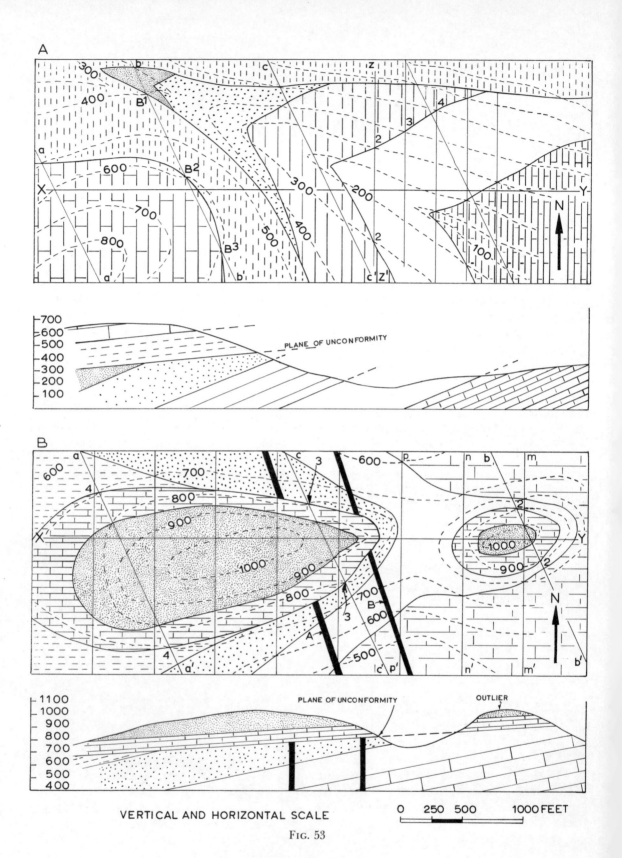

VERTICAL AND HORIZONTAL SCALE

FIG. 53

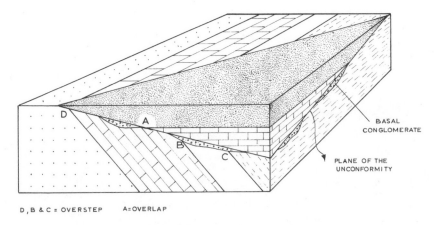

D, B & C = OVERSTEP   A = OVERLAP

DIAGRAM ILLUSTRATING UNCONFORMITY, OVERSTEP AND OVERLAP.

FIG. 54

Assynt Special Sheet displays numerous examples of this, where dykes in the Lewisian terminate abruptly against the overlying Torridonian.

Not all unconformities are recognisable at once, either on the map or in the field. It *is* possible for two superimposed series to have the same dip and strike as, for example, on the margin of the Thames Basin where the Tertiary and the underlying Chalk are so related. Such discontinuities and non-sequences require careful stratigraphical investigations for their ultimate recognition.

An unconformable series is commonly deposited in an advancing and enlarging water body. Such an advancing water body depositing sediment, will cover increasing areas with the progressively younger beds. Each successive younger member of a series will then extend beyond the boundary of the bed immediately below it: in such a situation the upper or newer bed is said to *overlap* the lower bed.

Since the lower (and usually older) member of an unconformable series has a greater dip than the upper, the latter will, in its forward advance, come to lie on different members of the lower series. The upper series is then said to *overstep* the members of the lower series. Sheet 265—Bath, shows the progressive overstep of the Lias on to the Old Red Sandstone, Carboniferous Limestone and Coal Measures. Figure 54 illustrates the three phenomena of unconformity, overstep and overlap.

Figures 53A and B are maps and sections embodying some of the criteria mentioned above for the recognition of unconformity.

Figure 53A shows the shale overstepping a number of different beds. The general pattern of the outcrops in the valley suggests a westerly dip in all beds. In examining this map to establish the presence of an unconformity, it is first necessary to determine the dip and strike of the beds. Consider the junction mudstone/marl: join points 2·2, this extended gives the strike line ZZ' which is the 300 ft strike line for this bed junction. Parallel lines to ZZ', through points 3 and 4 establish the strike as trending north to south and the dip as 1 in 2·5 westwards. The junctions oolitic limestone/marl: mudstone/coarse sandstone and coarse sandstone/sandstone can all be shown to be conformable with the mudstone/marl junction.

If now the junction of the chalk/shale be examined strike lines can be drawn through B^2 and B^3 and through a and a' at 600 ft and 500 ft respectively. This surface therefore strikes N26°W—N26°E and dips W26°S as 1 in 10. The base of the shale also has this strike and dip and is, therefore, the unconformable junction between the two series of beds. The section shows the unconformable relationships.

In Fig. 53B the base of the oolitic limestone oversteps the shale and the sandstone. Its dip and strike are established by joining points 2·2; 3·3; 4·4, thus marking the strike lines bb', cc', aa' at 900 ft, 800 ft and 700 ft on the base of the oolitic limestone: the dip of this surface, as a gradient, is 1 in 30 and the strike N26°W.

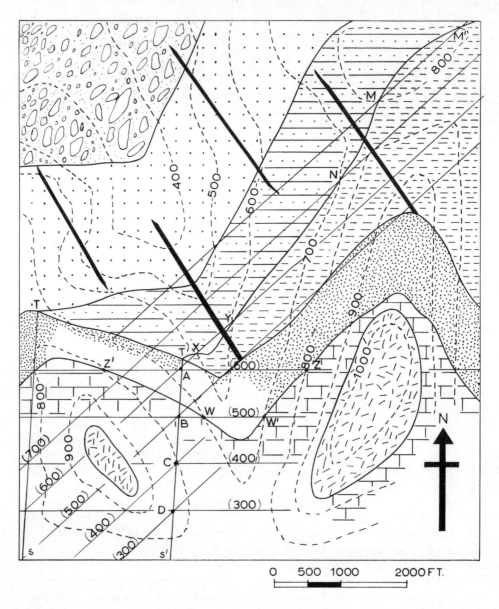

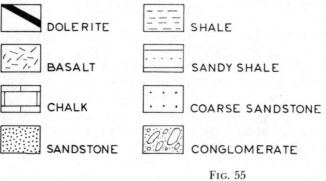

Fig. 55

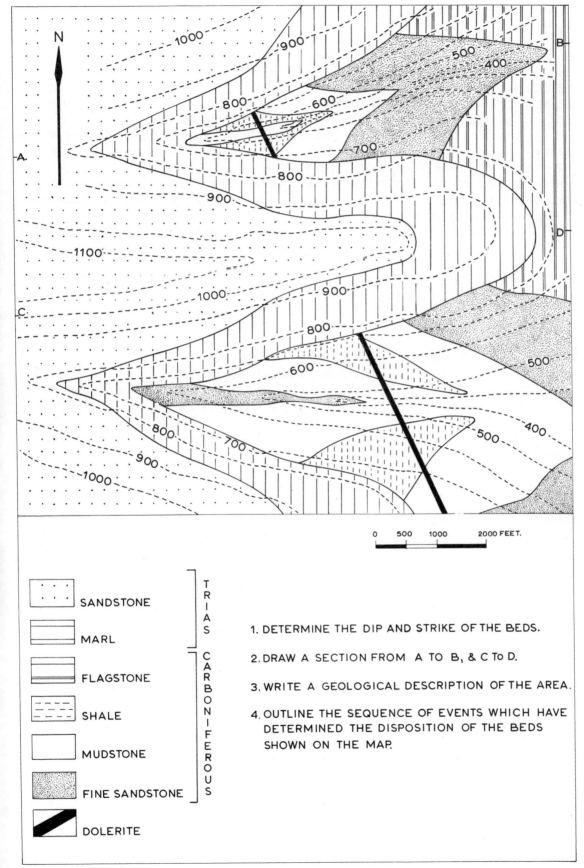

1. DETERMINE THE DIP AND STRIKE OF THE BEDS.
2. DRAW A SECTION FROM A TO B, & C TO D.
3. WRITE A GEOLOGICAL DESCRIPTION OF THE AREA.
4. OUTLINE THE SEQUENCE OF EVENTS WHICH HAVE DETERMINED THE DISPOSITION OF THE BEDS SHOWN ON THE MAP.

Fig. 56

Strike lines mm', nn' and pp' on the limestone/mudstone junction show it to dip westwards at 1 in 10. The mudstone/sandstone and sandstone/shale junctions have the same dip and strike.

There is discordance of dip and strike between two series of rocks in this map and the dykes are confined to the lower series of rocks, all being indicative of the unconformable relationship.

Notice the eastern outlier of the upper unconformable series, a structure to be described shortly.

The section clearly shows the relationship between the two series.

TO DETERMINE THE JUNCTION OF THE BEDS OF A LOWER SERIES WITH THE BASE OF THE UPPER SERIES IN AN UNCONFORMITY

In Fig. 55, two series of beds are in unconformable relationship with each other. This is shown by the transgression of outcrops across each other and by the restriction of the dykes to the lower series.

A strike line through XY gives the direction of strike of the lower series; other strike lines through MM' and N, parallel to XY, allow the amount and direction of dip to be calculated.

A line through ZZ' marks the strike of the upper series. The direction and amount of dip of this series can be determined by drawing a second strike line through WW'.

Suppose it is required to find the trace of the shale/sandy shale junction beneath the unconformity. Proceed as follows:

1. Draw the strike lines for the lower surface of the shale and on each line mark its values, viz. 600 ft.
2. Draw the strike lines for the lower surface of the sandstone of the upper series and number them clearly.
3. Mark the points of intersection of the strike lines of like value from each series (e.g. at A the two 600 ft strike lines intersect) and so indicate the junction of the two surfaces at that point. Similar reasoning will give the points B, C and D which, when joined together, mark the direction of the trace of the sandstone/sandy shale junction with the overlying sandstone, i.e. TS'.

If it is required to delimit the area of the upper series underlain by sandy shale a line TS, parallel to $T'S'$, would be the trace of the sandy shale/coarse sandstone junction: the area underlain by the sandy shale would be $TT'S'S$. Other areas could be delimited in this way.

The assumption is made that the unconformable junction is a plane surface.

Figure 56 provides an exercise in the examination of an unconformity.

PART 6

OUTLIERS AND INLIERS

The terms outlier and inlier refer to rock masses which have been isolated from their main outcrop either by erosion or tectonic disturbance: they are frequently associated with unconformity.

In Fig. 57 a gently dipping series of beds has been eroded leaving an isolated patch of sandstone in the east and, in the centre, a patch of sandstone and conglomerate. Both patches are relict portions of the main western outcrop of these rocks. In general, erosion is constantly uncovering the lower, older, members of any series so that the relict patches of the upper series come to be surrounded by the members of the lower, older series. These isolated outcrops are *outliers*. Examination of Sheet 286—Reigate, shows outliers of Lower Greensand resting on the older Wealden Beds.

Figure 57B depicts a lower folded series of beds overlain by an upper horizontal series. The strike lines drawn on the lower series reveal its folded nature: synclinal axes lie between the strike lines gg', ff' and between cc' and nn', whilst anticlinal axes are located between the strike lines dd' and aa' and between mm' and oo'. The trend of the outcrops of the upper series being parallel to the contours shows that the beds are horizontal. In this case erosion has developed a window in the upper series and exposed in that window an *inlier* of the older rocks.

Figure 58 portrays an inlier and an outlier produced through faulting. In Fig. 58A the strike lines aa', bb' and cc' indicate the direction and amount of dip of the upper series whilst strike lines mm', nn', pp', etc., give the direction of dip strike of the lower series. Two faults have isolated an oval area of the upper series and so produced a mass of the newer series surrounded by members of the older series; this is a *faulted outlier* such as is seen at Careg Cennen Castle, Sheet 230—Ammanford, where Carboniferous Limestone has been faulted against the Old Red Sandstone. In Fig. 58B a lower series of rocks, the dip and strike of which is seen from a study of the strike lines aa', bb', cc', etc., on the shale/sandstone junction, has been surrounded by a downfaulted horizontal sandstone; this structure is a *faulted inlier* such as the inlier of Silurian rocks at Usk shown on Sheets 249 and 233.

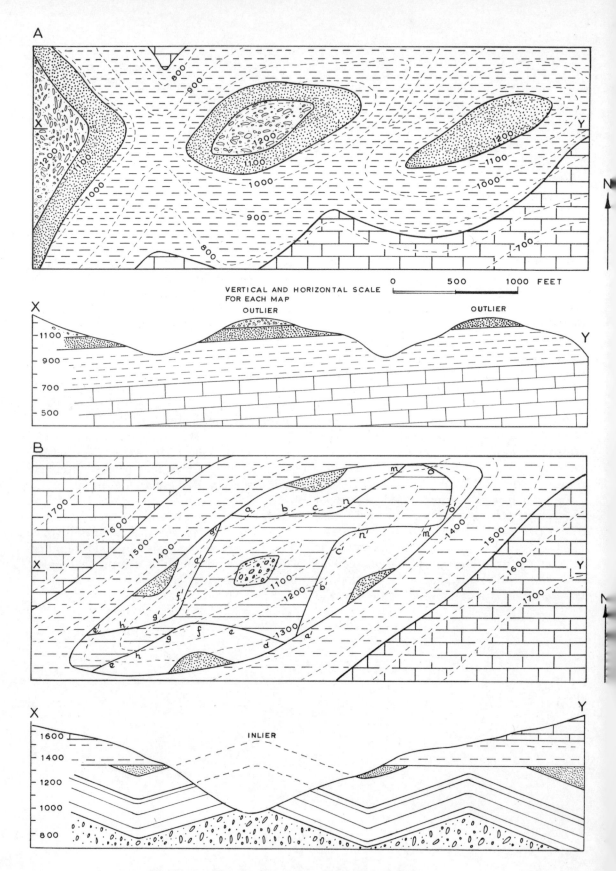

DIAGRAMS OF OUTLIERS AND INLIERS
Fig. 57

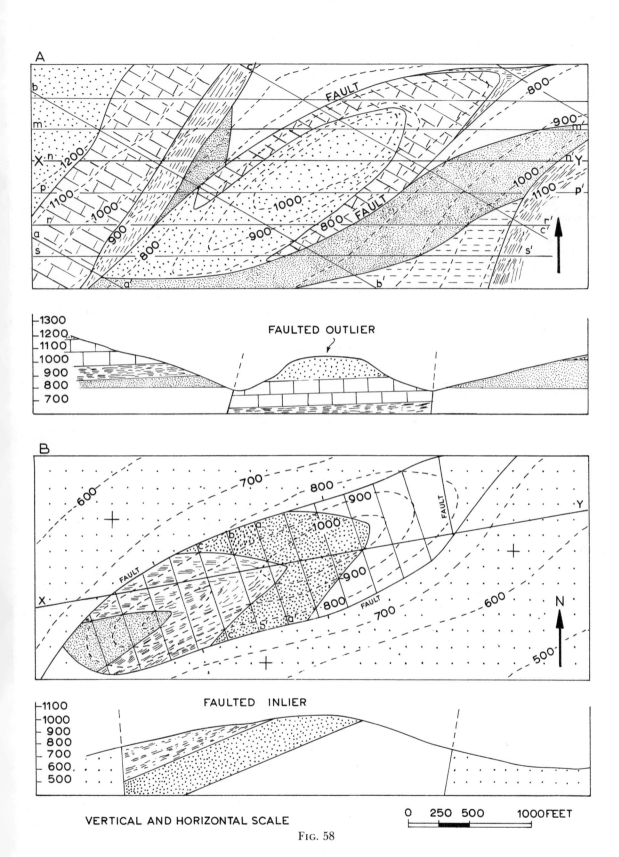

Fig. 58

PART 7

TO COMPLETE THE OUTCROPS OF BEDS FROM PARTIAL OUTCROPS

In mapping an area it is usual to observe only a few outcrops of rock boundaries from which the positions of the completed boundaries are calculated. The simple example, Fig. 59, is one way in which this completion of outcrops may be carried out.

The series of beds to be plotted on the map is a conformable one so that the strike and dip is the same for all the beds. An examination of the sandy shale/sandstone junction at D and C shows the junction to lie there at 500 ft O.D. whilst at A and B it lies at 400 ft O.D.: lines through these pairs of points are strike lines which establish the strike as east to west and the dip, as a gradient, 1 in 2 to the south. Since the beds are conformable the strike lines will be evenly spaced over the whole map. On the established dip there are four more strike lines to the north of DC and five more to the south of AB, for the sandy shale/sandstone junction. Because the beds are whole number multiples of one hundred feet in thickness the strike lines for the sandstone/sandy shale junction will also be strike line (of different value) for each of the other junctions. The strike line through F is the 500 ft strike line for the conglomerate/sandstone junction but is the 900 ft strike line for the top of the sandstone.

The outcrops of the various junctions between the beds can be drawn where the strike lines, with their values appropriate to the particular surface, intersect contours of like value. Here the junctions are indicated by long dashes.

TO PLOT ROCK OUTCROPS FROM BOREHOLE RECORDS

Problems involving the use of information from boreholes are often described as "three-point problems". The information required in order to plot such borehole information in terms of rock outcrops is:
1. The height of each borehole point of entry above sea level.
2. The vertical thickness of each bed present in each borehole.
3. Sufficient information between the three borehole records to allow the determination of the height of three points, one in each borehole, on one surface of the bed.

The problem is to plot the intersection of the inclined plane surface of the bed with the irregular topographic surface. The intersection of these two surfaces will only be visible where the topography and the surface of the bed occur at the same height above sea level.

Consider now Fig. 60, where three boreholes A, B and C yield information about the vertical thicknesses of beds occurring in them. The only common datum line between the three holes is the base of the limestone.

The topographic elevation at A is 1100 ft O.D. and the vertical thickness of the beds above the limestone is 1600 ft: the base of the limestone must there lie at −500 ft or 500 ft below sea level.

The topographic elevation at B is 850 ft O.D. and the vertical thickness of rock above the limestone base is 650 ft, hence the base of the limestone is at 200 ft O.D.

The topographic elevation of C is 500' O.D.: the limestone is 100' thick, hence the base of the limestone is at 400' O.D.

The dip and strike of the limestone base may now be calculated as follows:

1. Between B (200 ft O.D.) and A (−500 ft below sea level) the base of the limestone falls 700 ft. If BA be divided into seven equal parts then the distance between each division will be the horizontal distance in which the surface of the bed falls 100 ft, the elevation at each

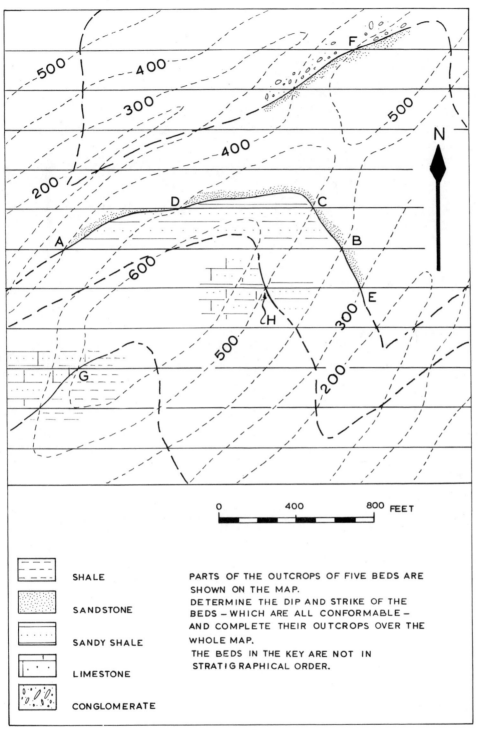

	SHALE
	SANDSTONE
	SANDY SHALE
	LIMESTONE
	CONGLOMERATE

PARTS OF THE OUTCROPS OF FIVE BEDS ARE SHOWN ON THE MAP.
DETERMINE THE DIP AND STRIKE OF THE BEDS – WHICH ARE ALL CONFORMABLE – AND COMPLETE THEIR OUTCROPS OVER THE WHOLE MAP.
THE BEDS IN THE KEY ARE NOT IN STRATIGRAPHICAL ORDER.

Fig. 59

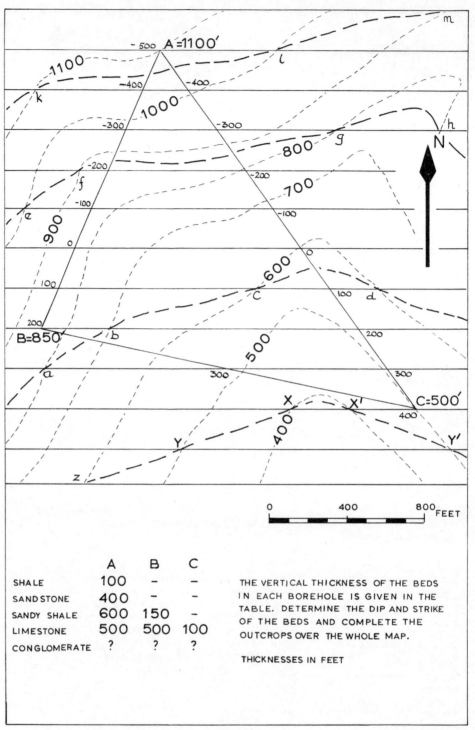

	A	B	C
SHALE	100	–	–
SANDSTONE	400	–	–
SANDY SHALE	600	150	–
LIMESTONE	500	500	100
CONGLOMERATE	?	?	?

THE VERTICAL THICKNESS OF THE BEDS IN EACH BOREHOLE IS GIVEN IN THE TABLE. DETERMINE THE DIP AND STRIKE OF THE BEDS AND COMPLETE THE OUTCROPS OVER THE WHOLE MAP.

THICKNESSES IN FEET

Fig. 60

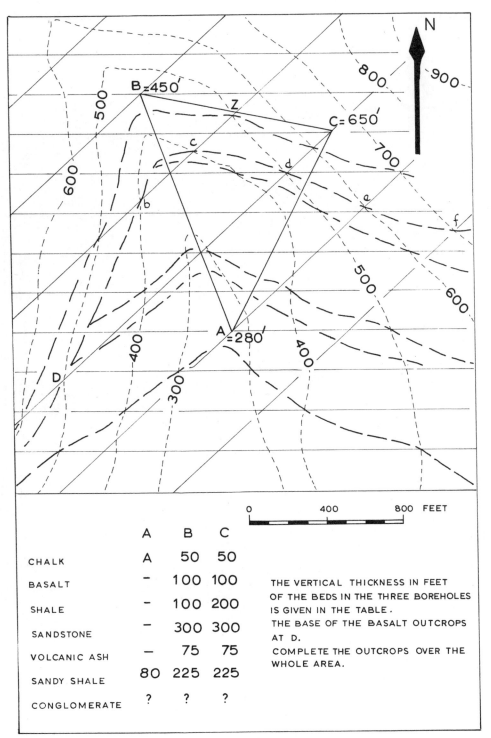

	A	B	C
CHALK	A	50	50
BASALT	–	100	100
SHALE	–	100	200
SANDSTONE	–	300	300
VOLCANIC ASH	–	75	75
SANDY SHALE	80	225	225
CONGLOMERATE	?	?	?

THE VERTICAL THICKNESS IN FEET OF THE BEDS IN THE THREE BOREHOLES IS GIVEN IN THE TABLE.
THE BASE OF THE BASALT OUTCROPS AT D.
COMPLETE THE OUTCROPS OVER THE WHOLE AREA.

Fig. 61

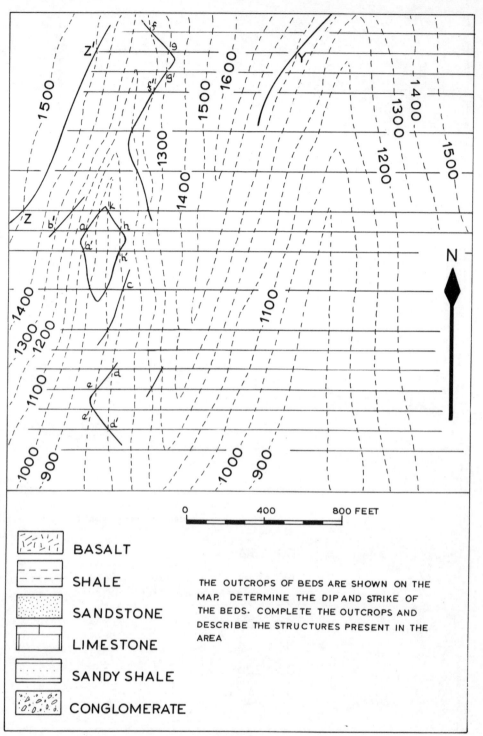

FIG. 62

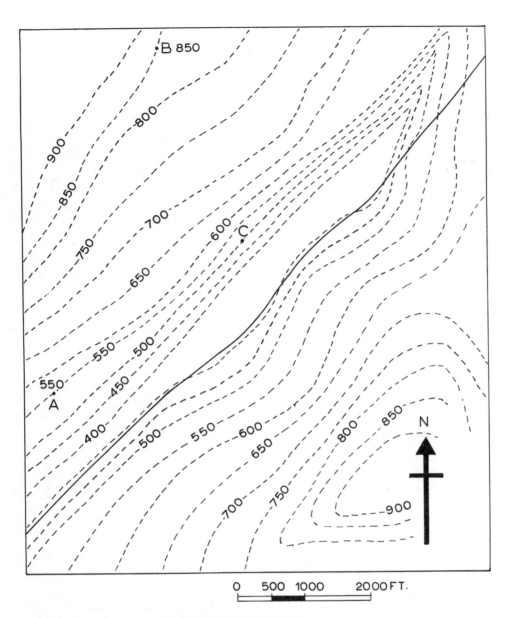

BOREHOLE	A 550' O.D.	B 850' O.D.	C 500' O.D.
LIMESTONE	50'	–	–
SANDY SHALE	400'	–	–
SHALE	50'	–	–
SANDSTONE	350'	150	50
MUDSTONE	300'	300	300
CONGLOMERATE	?	?	?

THE ABOVE TABLE SHOWS THE VERTICAL THICKNESS OF EACH BED PRESENT IN THE BOREHOLES A, B AND C.

1 DETERMINE THE DIP AND STRIKE OF THE BEDS IN THE WESTERN PART OF THE MAP, AND INSERT THE OUTCROPS OF THE BEDS.
2 THE FAULT HAS A DOWNTHROW OF 150' TO THE S.E. PLOT THE OUTCROPS OF THE BEDS S.E. OF THE FAULT, ASSUMING THAT THE DIP AND STRIKE ARE THE SAME ON EACH SIDE OF THE FAULT.

FIG. 63

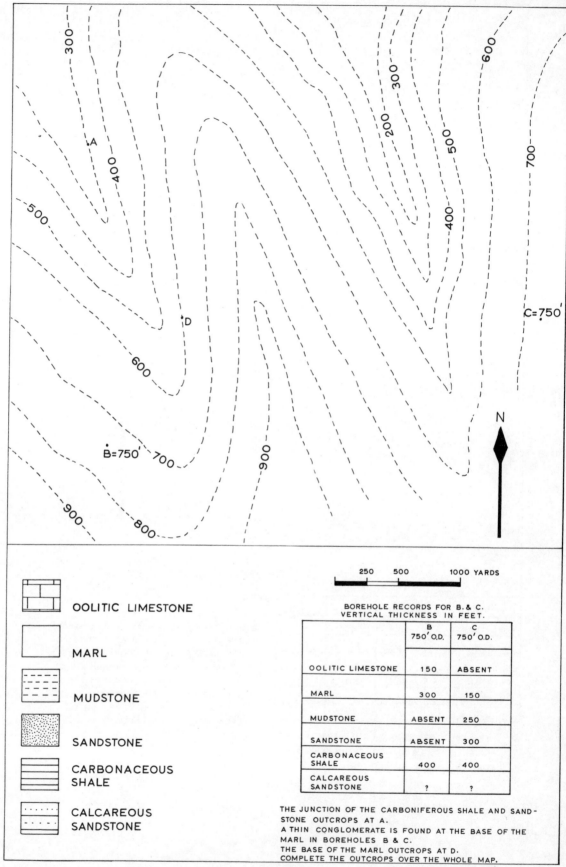

Fig. 64

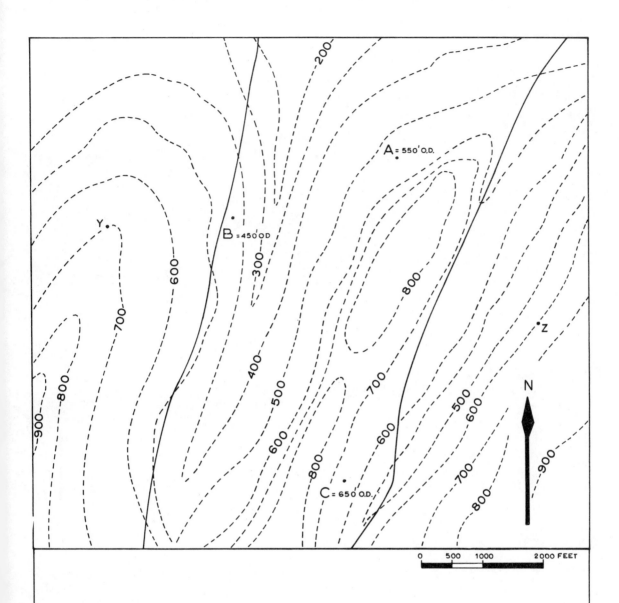

ROCK TYPE	A	B	C
SANDSTONE	-		400
LIMESTONE	-	250	400
SANDY SHALE	100	250	250
MUDSTONE	50	50	50
GRIT	300	300	300
SHALE	?	?	?

THE TABLE GIVES THE VERTICAL THICKNESS IN FEET OF THE DIFFERENT STRATA IN EACH OF THE BOREHOLES A, B AND C.

THE BASE OF THE MUDSTONE OUTCROPS AT Y.

THE BASE OF THE LIMESTONE OUTCROPS AT Z.

THE DIP AND STRIKE ARE CONSTANT OVER THE WHOLE AREA

DETERMINE THE DIP AND STRIKE OF THE BEDS, COMPLETE THE OUTCROPS OVER THE WHOLE AREA OF THE MAP, AND DETERMINE THE HADE AND THROW OF THE TWO FAULTS.

FIG. 65

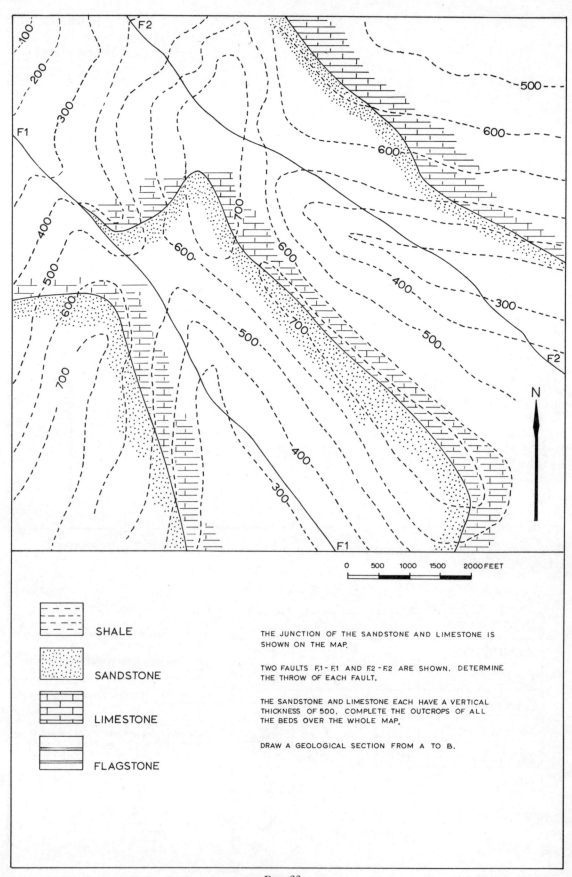

SHALE

SANDSTONE

LIMESTONE

FLAGSTONE

THE JUNCTION OF THE SANDSTONE AND LIMESTONE IS SHOWN ON THE MAP.

TWO FAULTS F.1 - F.1 AND F.2 - F.2 ARE SHOWN. DETERMINE THE THROW OF EACH FAULT.

THE SANDSTONE AND LIMESTONE EACH HAVE A VERTICAL THICKNESS OF 500. COMPLETE THE OUTCROPS OF ALL THE BEDS OVER THE WHOLE MAP.

DRAW A GEOLOGICAL SECTION FROM A TO B.

FIG. 66

point of subdivision should be marked as in Fig. 60.

2. Between C (400 ft O.D.) and A (-500 ft or 500 ft below sea level) the surface of the limestone falls 900 ft. Divide CA into 9 (nine) equal parts of which will be the horizontal equivalent for a fall or 100 ft along CA. Mark the topographic height of each point.

3. Between C (400 ft O.D.) and B (200 ft O.D.) the base of the limestone falls 200 ft. Divide CD into two equal parts: each part is the horizontal equivalent of a fall of 100 ft in the direction CB. Mark the topographic height of the divisions.

Now ABC is part of the plane surface of the base of the limestone and contours traced upon it will be straight lines; these straight lines are here obtained by joining the pairs of points of like elevation. When drawn across the whole map they are the strike lines at 100 ft intervals for the base of the limestone. Should there be portions of the map where the consideration of ABC has not caused strike lines to be drawn then, as in the south of the present map further strike lines should be drawn at the intervals calculated in ABC. When the strike lines have been drawn the strike of the base of the limestone is proved to be east/west and the dip is due north at gradient of 1 in 2. The outcrop of the base of the limestone may now be traced. As already shown the outcrop of a bed is only seen where its height above sea level coincides with the topographic height above sea level. In Fig. 60 the 400 ft strike line intersects the 400 ft contour at X and X'; the 500 ft strike line intersects the 500 ft contour at Y and Y' and the 600 ft strike line intersects the 600 ft contour at Z: the trace of the outcrop of the base of the limestone, which is the same as the conglomerate/limestone junction is $ZYXX'Y'$.

The borehole at A shows the vertical thickness of the beds present in the area. The limestone has a vertical thickness of 500 ft: this being so and the bed being a whole number of hundreds of feet in thickness the 400 ft strike line on X-X' for the base of the limestone is the 900 ft strike line for the upper surface. If 500 be added to each of the strike lines for the base of the limestone, then their intersections with the contours at a, b, c, and d enable the outcrop of the top of the limestone to be plotted.

If now to each strike line value there be added a cumulative succession the vertical thicknesses of the beds present then the sandy shale/sandstone junction appears at e, f, g, h and the sandstone/shale junction at k, l, m.

As a second example of the completion of outcrops from borehole records consider Fig. 61. In the three boreholes A, B and C the vertical thicknesses of the strata are recorded. If the base of the sandy shale is taken as the datum plane in each borehole then the beds up to the base of the shale can be plotted as in Fig. 60. The beds are seen to have an east/west strike and a southerly dip with a gradient of 1 in 2. The thickness of the shale is different in the two boreholes B and C and there is insufficient evidence to enable the top of the shale to be plotted. Three points of elevation on the base of the basalt are obtainable: at C it occurs at 500 ft O.D. and at B at 300 ft O.D.; the base also outcrops at D at 500 ft O.D. A line joining CD is the 500 ft strike line for the base of the basalt. The dip and strike of the basalt is obtained by joining CB and D: between C and B the basalt base falls 200 ft from 500 ft to 300 ft, the mid point of CB (Z) will lie at 400 ft O.D. and a line through Z parallel to CD is the 300 ft strike line for the basalt. Other strike lines drawn across the map spaced as those already determined enable the trace base of the basalt $Dbcdef$ to be plotted. The chalk is shown to lie conformably on the basalt but the basalt rests unconformably on the lower series from the conglomerate to the shale.

Figure 62 is an exercise in the completion of outcrops with an unconformable upper series and an underlying folded series. The strike of the upper series is given by $Z'Y$, and the dip may be determined by drawing a parallel strike line through Z.

Strike lines through ah and $a'h'$ give the strike of the lower series. If all the strike lines are drawn the folds are seen to be asymmetrical.

Figures 63, 64, 65 and 66 are exercises in the completion of outcrops.

APPENDIX I

THE ANGLE OF DIP IN EXAGGERATED VERTICAL SCALE IN GEOLOGICAL SECTIONS

When geological sections are constructed the true dip will be shown only in those sections in which the vertical and horizontal scales are the same. If the vertical scale is exaggerated, as is common practice in drawing sections across the 1 inch to 1 mile Geological Survey Maps, then the dip must be increased to fit the new scale. Figure 67 shows how the dip increases when the vertical scale is increased by two and three whilst the horizontal scale remains constant. Let AB and FC be drawn on the same scale and let there be a dip of 20° from B. If the vertical scale is doubled, i.e. to FD, then the angle increases to 36°3′ whilst increasing the vertical scale three times to FE produces a dip of 47°31′. The table shows these increases to be predictable for the tangent of 36°3′ is 0·7280 which is twice the tangent of 20°, the true dip and the tangent of 47°31′ is 1·0820 which is three times the tangent value of 20°.

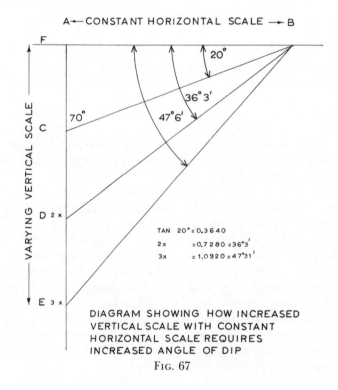

DIAGRAM SHOWING HOW INCREASED VERTICAL SCALE WITH CONSTANT HORIZONTAL SCALE REQUIRES INCREASED ANGLE OF DIP

FIG. 67

APPENDIX II

Problem 1 (Fig. 68). If two apparent dips in a bed of sandstone are 54°3′S 50°W and 63°37′S 20°E, determine the amount and direction of true dip.

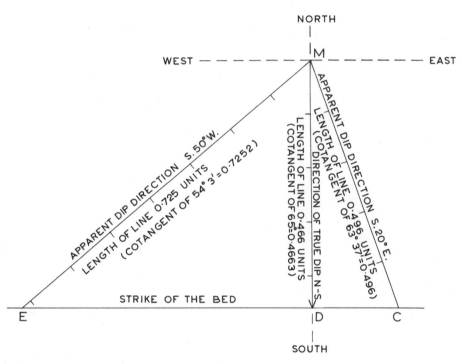

Fig. 68

1. Set out the north/south and east/west directions which intersect at M.
2. From M draw ME in a direction S50°W, and MC in a direction S20°E.
3. Use a suitable scale and draw $ME = 0.7252$ units long and $MC = 0.4960$ units long. These distances are a representation of the cotangent values of 54°3′ and 63°37′.
4. Join EC; EC is the direction of strike of the bed of sandstone.
5. From M drop a perpendicular to meet EC at D. MD represents the direction of true dip. If $E\hat{M}D$ is measured with a protractor it will be found to be 50°, so the *direction* of true dip is due south.
6. MD measured in the same units as those in which ME and MC were drawn is found to be 0·466 units in length. This represents the value of the cotangent of the true dip, which, from tables, is found to be 65°.

A second graphical method (Fig. 69) of solving this problem is as follows:
1. Lay out the north/south and east/west directions and let them intersect at O.
2. From O draw OZ in the direction S50°W and OY in the direction S20°E.
3. At O draw a perpendicular to OZ and to OY.

69

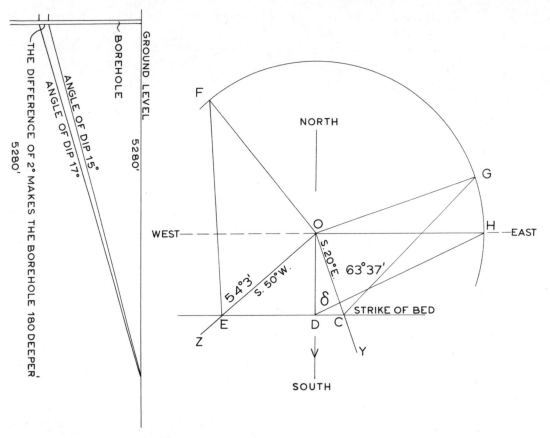

Fig. 69

4. With any convenient radius draw an arc to cut these two perpendiculars at F and G respectively.
5. At F lay off a line to cut OZ at E so that OFÊ is the complement of the apparent dip angle (90°–54°3′): OÊF is the angle of apparent dip.
6. At G lay off a line to cut OY at C so that OĜC is the complement of the apparent dip angle (90°–63°37′): OĈG is the angle of apparent dip.
7. Join E and C: the line EC is the direction of strike of the bed.
8. From O draw OD perpendicular to EC: OD is the direction of true dip.
9. From O draw a perpendicular to OD to cut the arc at H.
10. Join DH: the angle OD̂H is the angle of true dip and measures 65°.

The above problem may also be solved trigonometrically for a relationship exists between true and apparent dip.

Let δ = the angle of true dip,
α = the angle of apparent dip,
θ = the angle between the direction of true and apparent dips.

The relationship between the three is expressed as follows:
$$\tan \delta = \frac{\tan \alpha}{\cos \theta}$$

This formula could not be applied to the above problem directly since θ is unknown. However, if the two apparent dips are known and the angle between their directions is known then θ can be calculated as shown below:

Consider Fig. 73 in which:

AB is unity,
$EB = \cot \alpha$,
$CB = \cot \beta$,
$A\hat{D}B$ (δ) = the angle of true dip,
$E\hat{B}C$ (ψ) = the angle between the angles of apparent dip,
$A\hat{D}B$ (θ) = the angle between the direction

70

of true dip and the direction of the apparent dip, α.

$D\hat{B}C$ $(\psi - \theta)$ = the angle between the direction of true dip and the direction of the apparent dip, β.

Then $BD = CB \cos(\psi - \theta)$
$BD = EB \cos \theta$

since $EB = \cot \alpha$ and $CB = \cot \beta$

$\cot \alpha \cos \theta \quad \cot \beta \cos(\psi - \theta)$
$\cos \alpha \cos \theta = \cot \beta (\cos \psi \cos \theta + \sin \psi \sin \theta)$

$\cot \alpha \tan \beta \cos \theta = \cos \psi \cos \theta + \sin \psi \sin \theta$
$\cot \alpha \tan \beta \cos \theta - \cos \psi \cos \theta = \sin \psi \sin \theta$

$$\frac{\cot \alpha \tan \beta - \cos \psi}{\sin \psi} = \frac{\sin \theta}{\cos \theta}$$

i.e. $\tan \theta = \dfrac{\cot \alpha \tan \beta - \cos \psi}{\sin \psi}$

or $\tan \theta = \mathrm{cosec}\, \psi\, (\cot \alpha \tan \beta - \cos \psi)$

Referring now to the problem above

$\tan \theta = \mathrm{cosec}\, 70\, (\cot 54°3' \tan 63°37' - \cos 70)$

$= 1·0642\, (0·7252 \times 2·016 - 0·3420)$

$= 1·0642 \times 1·120$

$\tan \theta = 1·1919$

from tables $\theta = 50°$

Now using the formula for the relation between the true and apparent dip, the true dip can be found.

$$\tan \delta = \frac{\tan \alpha}{\cos \theta}$$

$$= \frac{\tan 54°3'}{\cos 50°}$$

$$= \frac{1·379}{0·6428}$$

$$= 2·145$$

from tables $\theta = 65°$

Problem 2 (Fig. 70). When the amount and direction of true dip are given, to determine the apparent dip in any other direction.

In Fig. 70 let A-B be the direction of a true dip of 25°: it is required to find the dip in the direction A-D.

Through A draw FAE at right angles to A-B. The line FAE is the strike of the bed dipping in the direction AB.

Along AF mark any convenient length, say AC, and construct an angle ACM equal to the

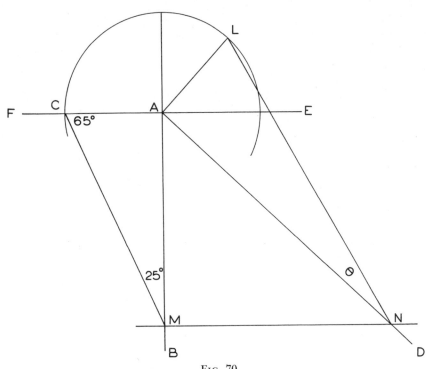

Fig. 70

complement of the angle of dip (90° − 25° = 65°).

Through *M* draw *MN* parallel to *FAE* to cut *AD* at *N*.

Draw *AL* at right angles to *AD* at *A* making *AL* equal in length to *AC*.

Join *LN*.

The angle of apparent dip in the direction *AD* is $A\hat{N}L$ (θ).

Problem 3 (Fig. 71). If the direction and amount of true dip are given to find the direction in which a bed will have a given dip. (Such dips will always be less than the true dip.)

At *N* construct the angle $A\hat{N}E$ equal to the complement of the required dip, i.e. 90° − 35° = 55°.

From *N* extend the line enclosing the angle to cut *DE* at *F*.

With *A* as centre and a radius equal to *AF* draw an arc to cut *JCK* at *G* and *H*.

Join *H* and *G*.

The directions *AH* and *AG* are the required directions in which the dip is 35°. (Note that there will always be two directions on a dipping bed where the apparent dip is the same value.)

In order to indicate the practical application of the problems set out above the following

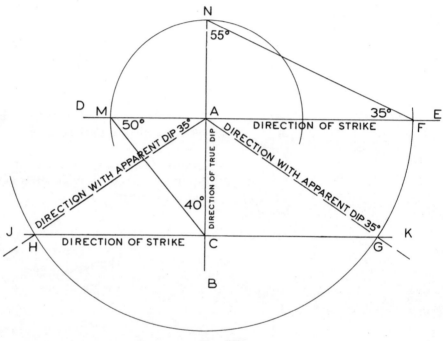

Fig. 71

In Fig. 71 let *AC* be the direction of true dip and let that dip be 40°. It is required to find the direction in which the dip is 35°.

Draw *AC* in the direction of true dip and, through *A*, draw *MAE* which is the direction of strike for the dip along *AC*.

On *AD* mark off *AM* any suitable length.

At *M* construct an angle $A\hat{M}C$ equal to the complement of the angle of full dip, i.e. 90° − 40° = 50°.

Through *C* draw *JCK* parallel to *MAE*.

Produce *CA* to *N* making *AN* equal in length to *AM*.

exercise is examined.

At a proposed colliery site – Fig. 72A – it is required to find:

1. The dip and strike of the coal seam which outcrops in the north-west of the site.
2. The depth to which an airshaft *B* must be sunk to meet the coal seam at *A*.
3. The direction from the bottom of the pit shaft *C* in which a roadway would have a gradient of 1 in 10.

1. In Fig. 72A the coal seam intersects the 200 ft contour at *b* and *b'*, a line joining these

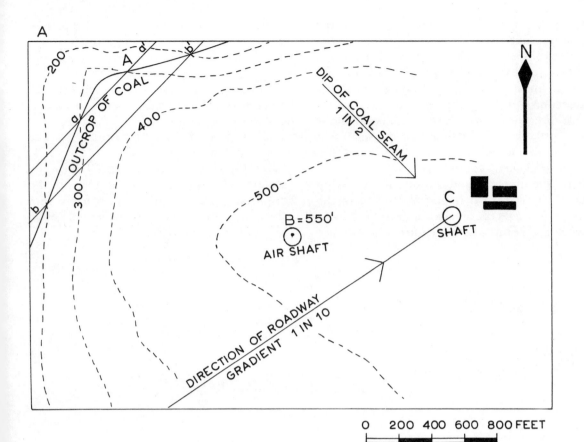

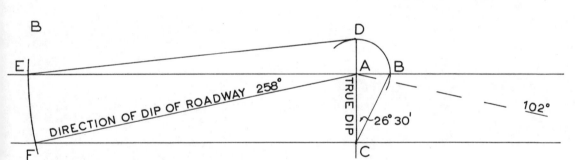

CALCULATION OF THE DIRECTION IN WHICH THE ROADWAY FROM THE PIT BOTTOM RISES AT 1 IN 10.

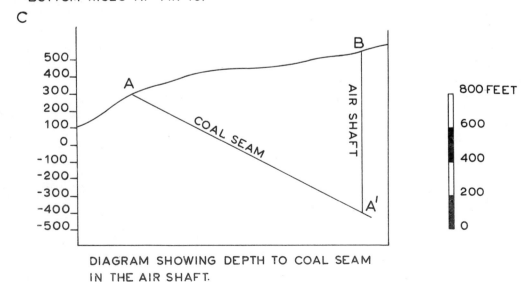

DIAGRAM SHOWING DEPTH TO COAL SEAM IN THE AIR SHAFT.

FIG. 72

two points gives the direction of strike. If the points a and A, where the seam intersects the 300 ft contour are joined, a second strike line is determined. A line at right angles to the two strike lines (or stratum contours) gives the direction of the dip the seam falls in a south-east direction (135°). The distance between the two strike lines is 200 ft so that the seam falls 100 ft in a horizontal distance of 200 ft and thus has a gradient of 1 in 2 (26°34′).

2. The direction and amount of dip having been determined it is possible to obtain the depth to which the air shaft B must be sunk to meet the seam at A'. Measure the distance A-B, which is in the direction of full dip. This distance is 1400 ft and, with a gradient of 1 in 2 the seam will fall 700 ft in that distance and at A' the seam will be at an elevation of −400 ft. Since the top of the air shaft is at 550 ft O.D. the depth from the surface to the point where the shaft meets the seam is 950 ft (Fig. 72C).

3. This part of the investigation can be carried out in the following way:

Using the construction in Fig. 71 it can be seen in Fig. 72B that two roadways with the required gradient may be constructed from the bottom of the shaft C, one in a direction 258° and the other at 102°.

APPENDIX III

RELATIONSHIP BETWEEN TRUE AND APPARENT DIP

The relationship between the true and apparent dip is expressed as:

$$\tan \delta = \frac{\tan \alpha}{\cos \theta}$$

where δ = the angle of true dip,
 α = the angle of apparent dip,
 θ = the angle between the direction of true and apparent dip.

From Fig. 73:
Let AB = unity,
then $BD = \cot \delta$ and $EB = \cot \alpha$,
and since $BDE = 90°$,
then $BD = EB \cos \theta$,
$\therefore \cot \delta = \cot \alpha \cos \theta$
or $\tan \delta = \dfrac{\tan \alpha}{\cos \theta}$.

If two apparent dips are known and the angle between them is given then θ can be calculated as follows:

Refer to Fig 73

AB = unity.
Then $EB = \cot \alpha$,
$CB = \cot \beta$,
$A\hat{D}B$ (δ) = the angle of true dip,
$E\hat{B}C$ (ψ) = the angle between the two angles of apparent dip,
$E\hat{D}B$ (θ) = the angle between the direction of true dip and the direction of the apparent dip, α.
$D\hat{B}C$ ($\psi - \theta$) = the angle between the direction of true dip δ and that of the apparent dip, β.

Then $BD = CB \cos (\psi - \theta)$
$BD = EB \cos \theta$

and since $EB = \cot \alpha$ and $CB = \cot \beta$

then $\cot \beta \cos (\psi - \theta) = \cot \alpha \cos \theta$.

Hence $\cot \alpha \cos \theta = \cot \beta (\cos \psi \cos \theta + \sin \psi \sin \theta)$

$\cot \alpha \tan \beta \cos \theta = \cos \psi \cos \theta + \sin \psi \sin \theta$

$\cot \alpha \tan \beta \cos \theta - \cos \psi \cos \theta = \sin \psi \sin \theta$

$$\frac{\cot \alpha \tan \beta - \cos \psi}{\sin \psi} = \frac{\sin \theta}{\cos \theta} = \tan \theta$$

i.e. $\tan \theta = \dfrac{\cot \alpha \tan \beta - \cos \psi}{\sin \psi}$
 $= \operatorname{cosec} \psi (\cot \alpha \tan \beta - \cos \psi)$

Referring now to a specific problem.

Two apparent dips are 54°3′ in a direction S50°W and 63°37′ in a direction S20°E: determine the amount and direction of true dip.

Determine θ the angle between the direction of apparent dip in the direction S50°W and the direction of true dip.

$\tan \theta = \operatorname{cosec} 70° (\cot 54°3′ \tan 63°37′ - \cos 70°)$
$= 1·0642 (0·7252 \times 2·016 - 0·3420)$
$= 1·0642 \times 1·120$
$= 1·1919$

from tables $\theta = 50°$ and the direction of true dip is therefore due south.

The true dip is found from the formula

$$\tan \delta = \frac{\tan \alpha}{\cos \theta}$$

where δ = angle of true dip,
 α = angle of apparent dip,
 θ = angle between the two dip directions.

Hence $\tan \delta = \dfrac{\tan 54°3′}{\cos 50°}$
$= \dfrac{1·379}{0·6428}$
$= 2·145$

from tables $\theta = 65°$

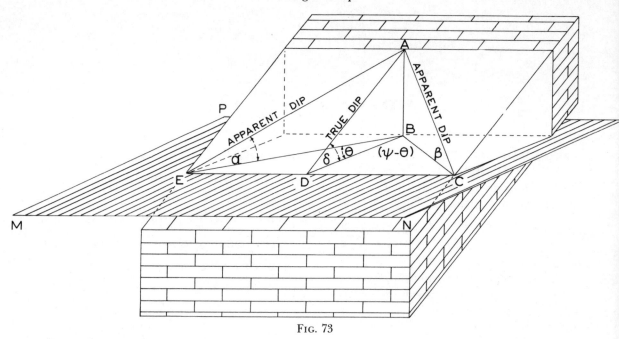

Fig. 73

The answer to the problem is that the true dip of the bed is 65° due south.

It is often difficult to be certain of the value of a single dip measurement made in the field and it is better to take several dips, to measure the angle between those directions and then calculate the dip from that information.

N.B. It has been assumed throughout that directions of apparent dip have been measured in azimuth.

APPENDIX IV

THE SOLUTION OF DIP AND APPARENT DIP PROBLEMS USING THE STEREOGRAPHIC PROJECTION

It is important in the study of a geological map to be able to visualise the three-dimensional picture it portrays. Since the sedimentary rocks, characteristically occurring in sheets, occupy so large a portion of the land surface and so many metamorphic and igneous rocks also have a sheet-like form, the problem of interpreting rock structures is frequently one of understanding the disposition and relationship of such sheets with each other and their intersection with the surface of the earth.

The common method used by structural geologists to record such rock relationships is the stereographic projection. This device can be understood by considering the lines of longitude, familiar on geographical globes. Imagine a glass sphere on which the lines of longitude are drawn through the north-south poles. Let the sphere be cut in half along a plane passing through the north and south poles to let an observer view the lines of longitude on the hemisphere from the extremity of a diameter of the sphere normal to the north/south plane (Fig. 74A). The lines of longitude on the outer surface of the hemisphere would appear as curved lines on that plane and these would be their stereographic projections (Fig. 74B).

If the lines of longitude are regarded as the traces of the intersections of planes passing through the centre of the sphere with its surface then, the curved lines, seen on the north/south medium plane, mark the projections of those traces on to that plane. Such traces could equally well be those of sheets of rock, the curved lines indicating varying dips and the diameter the strike. Countless families of planes of this type can be constructed about any diameters so that any attitude of rock sheet can be shown in this type of projection.

The two problems worked out below will indicate how the projection is used in the solution of dip and strike problems, the simplest examples of its use.

The stereographic net most used in the solution of dip and strike problems is the Wulff net shown in Fig. 75. This net consists of the projection of great circles (great circles are the traces of planes which pass through the centre of a sphere) about a principal axis and at right angles to it.*

The great circles may be projected as follows. Imagine a plane dipping towards the east at 50° as in Fig. 74A. If such a plane intersects the surface of a sphere and passes through its centre, the intersection with that body will be a great circle; such an intersection is seen in Fig. 74A intersecting the hemisphere along the arc *abc*. The stereographic projection of this is obtained by projecting from the arc of intersection on the sphere points such as *a*, *b* and *c* to the zenithal point – Fig. 74 – EYE and cutting the medial plane at *a'*, *b'* and *c'*: with sufficient points projected the stereographic representation of the plane dipping east at 50° may be drawn. As shown in Fig. 76 small circles are also projected and these are analogous to the lines of latitude seen on map projections. Both great and small circles are marked at 2° intervals and the 10° intervals are marked by thicker lines.

In the problems discussed below a piece of tracing paper overlaying the net and pinned through its centre is used to draw the necessary constructions and is easily rotated around the centre pin.

It should be noted that this projection is 'angle true'.

Problem 1

To determine the true dip and strike of a bed given the direction and amount of two apparent dips. Suppose the apparent dips are

---

*It should be understood that in structural geology the *lower* hemisphere is used, but in using this same projection in crystallography the upper hemisphere is used. This is because in general the upper and lower halves of the crystal are identical.

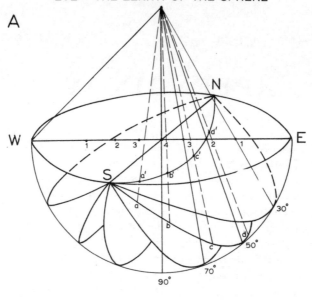
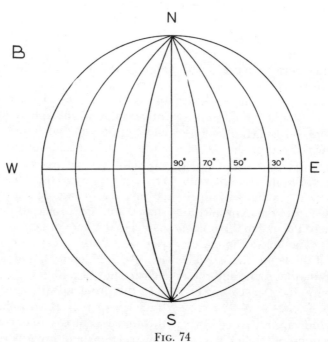

Fig. 74

$A = 49°$ in a direction 145° (S25°E) and $B = 40°$ in a direction 240° (S60°W) (Fig. 76B).

1. Mark on the tracing paper the E-W and N-S directions.
2. Mark on the tracing paper the directions of the apparent dips on the equatorial circle at $A = 145°$ and $B = 240°$ (Fig. 76B).
3. Rotate the tracing paper so that A lies over the south pole of the principal axis; mark off the dip of 49° along the semi-axis S-O. This is the point C (Fig. 76B).
4. Rotate the tracing paper so that B lies on the south pole of the principal axis; mark off the dip of 40° along the semi-axis S-O. This is the point D (Fig. 76C).

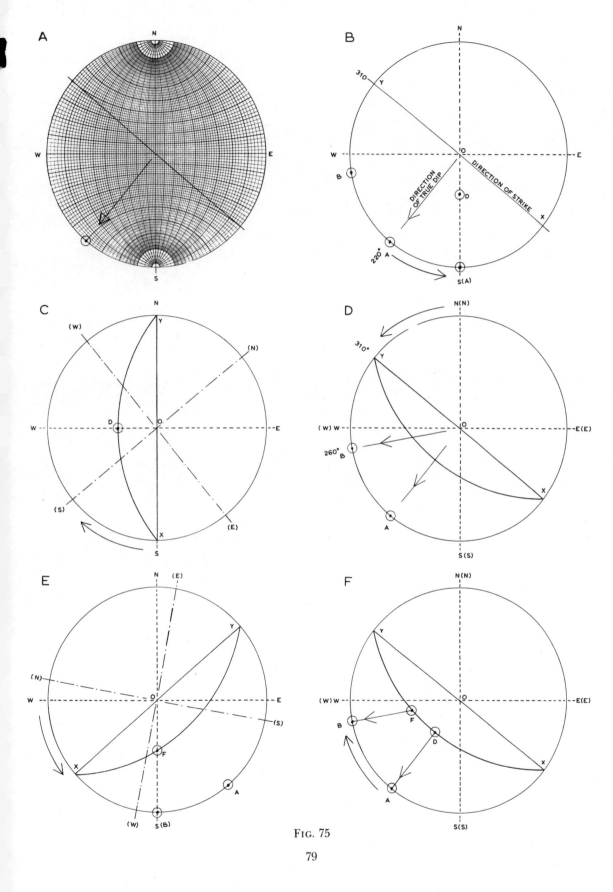

Fig. 75

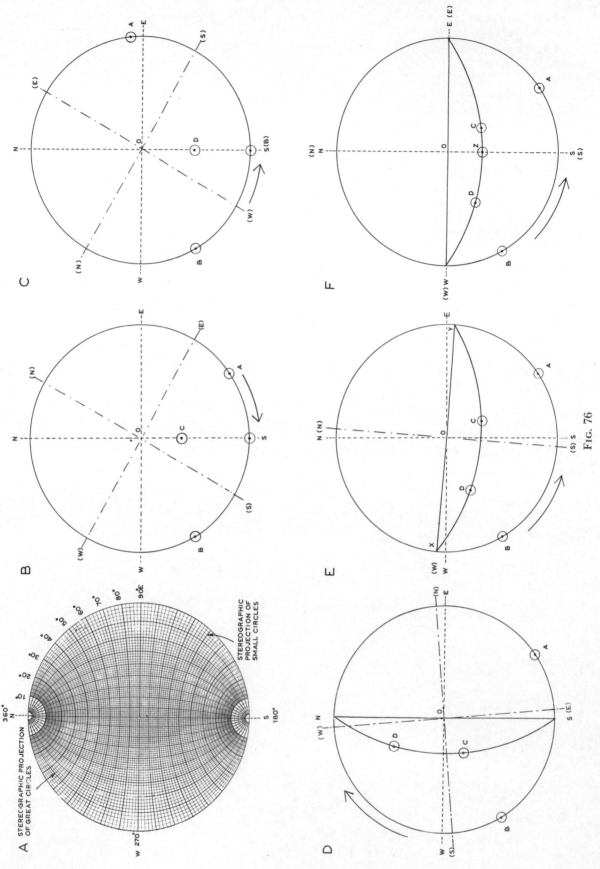

Fig. 76

5. Rotate the tracing paper until C and D lie on the same great circle. Line in on the tracing paper the great circle on which C and D lie (Fig. 76D).
6. Rotate the tracing paper until A and B occupy their original positions as in Fig. 76B (Fig. 76E). XOY is now the direction of strike of the bed and the direction of dip is along (S)-O, i.e. 185° (S5°W).
7. Rotate the tracing paper so that the dip direction lies along the principal axis S-O. The intersection of the great circle, on which D and C lie, with the principal axis at Z enables the dip to be measured as 56° (Fig. 76F).

The true dip of the bed on which the two apparent dips were given is 56° in a direction 185° (S59°W).

Problem 2

A bed of rock has a true dip of 50° in a direction 220°: determine the dip in a direction 260°.
1. Lay on the Wolff net a piece of tracing paper as in Problem 1, and mark E-W and N-S (Fig. 75B).
2. Rotate the tracing paper so that the strike direction X-Y coincides with the N-S direction in Fig. 75C.
3. With the strike direction X-Y parallel to the N-S direction mark off the dip of 50° along W-O: this is the point D (Fig. 75C).
4. On the tracing paper draw in the great circle on which D lies (Fig. 75C).
5. Rotate the tracing paper so that the strike lies in its true position 310° (Fig. 75D).
6. Mark on the tracing paper and in the equatorial plane the point B which is the direction 260° in which the dip is required (Fig. 75D).
7. Rotate the tracing paper until D lies above the south pole of the principal axis which is cut by the great circle on which D lies at F: the angle of dip is then read along the axis S-O as 42° (Fig. 75E).
8. Rotate the tracing paper so that the strike X-Y lies in its original position at 310°: the direction of the true and apparent dips are now shown in their true position, the true dip being 50° in a direction 220° and the apparent dip in the direction 260° being 42° (Fig. 75F).

APPENDIX V

MISCELLANEOUS EXAMPLES

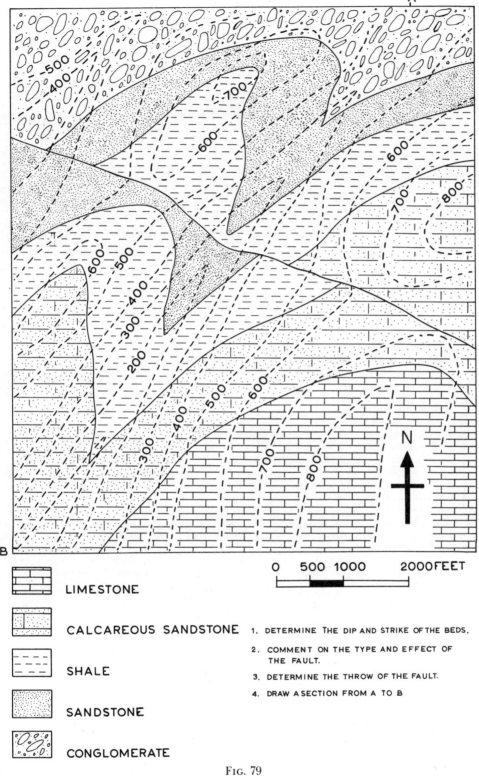

	LIMESTONE
	CALCAREOUS SANDSTONE
	SHALE
	SANDSTONE
	CONGLOMERATE

1. DETERMINE THE DIP AND STRIKE OF THE BEDS.
2. COMMENT ON THE TYPE AND EFFECT OF THE FAULT.
3. DETERMINE THE THROW OF THE FAULT.
4. DRAW A SECTION FROM A TO B

FIG. 79

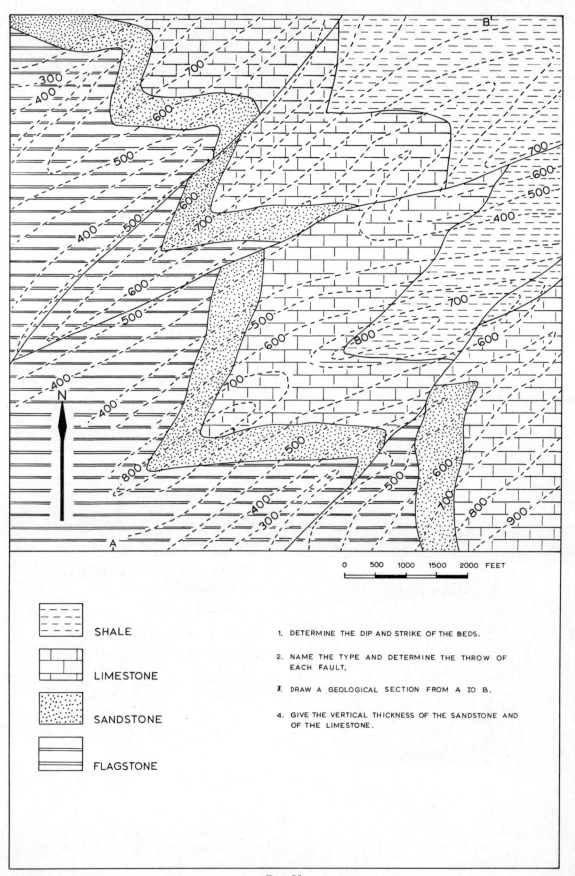

Fig. 80

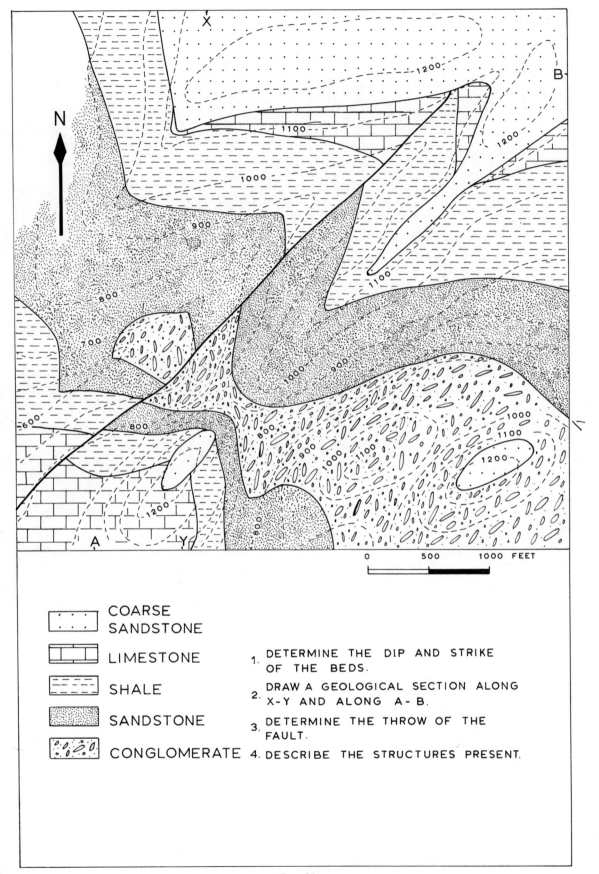

FIG. 81

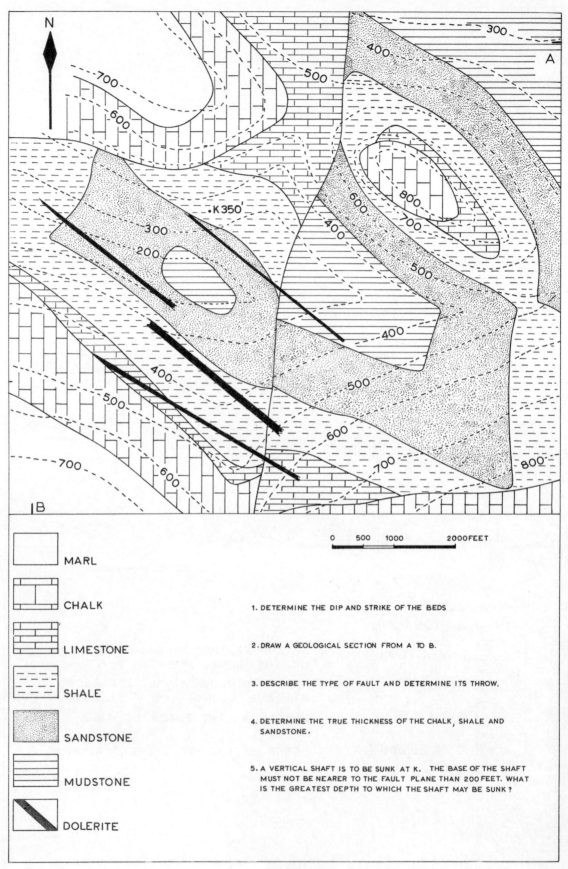

MARL

CHALK

LIMESTONE

SHALE

SANDSTONE

MUDSTONE

DOLERITE

1. DETERMINE THE DIP AND STRIKE OF THE BEDS

2. DRAW A GEOLOGICAL SECTION FROM A TO B.

3. DESCRIBE THE TYPE OF FAULT AND DETERMINE ITS THROW.

4. DETERMINE THE TRUE THICKNESS OF THE CHALK, SHALE AND SANDSTONE.

5. A VERTICAL SHAFT IS TO BE SUNK AT K. THE BASE OF THE SHAFT MUST NOT BE NEARER TO THE FAULT PLANE THAN 200 FEET. WHAT IS THE GREATEST DEPTH TO WHICH THE SHAFT MAY BE SUNK?

Fig. 82

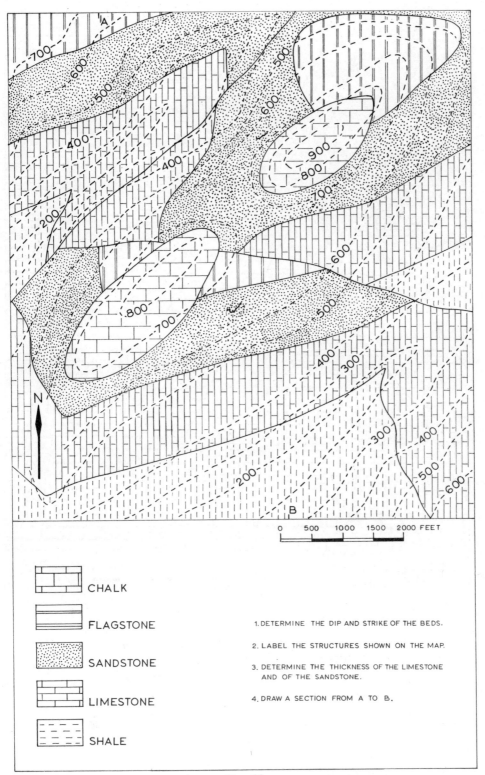

Fig. 83

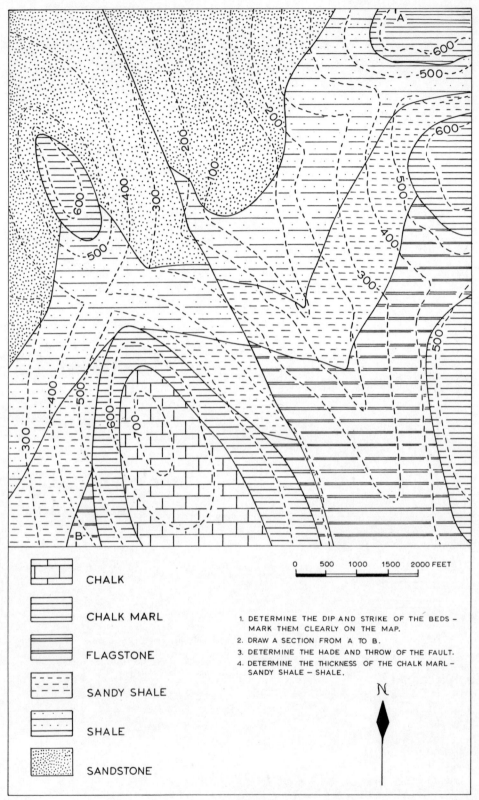

Fig. 84

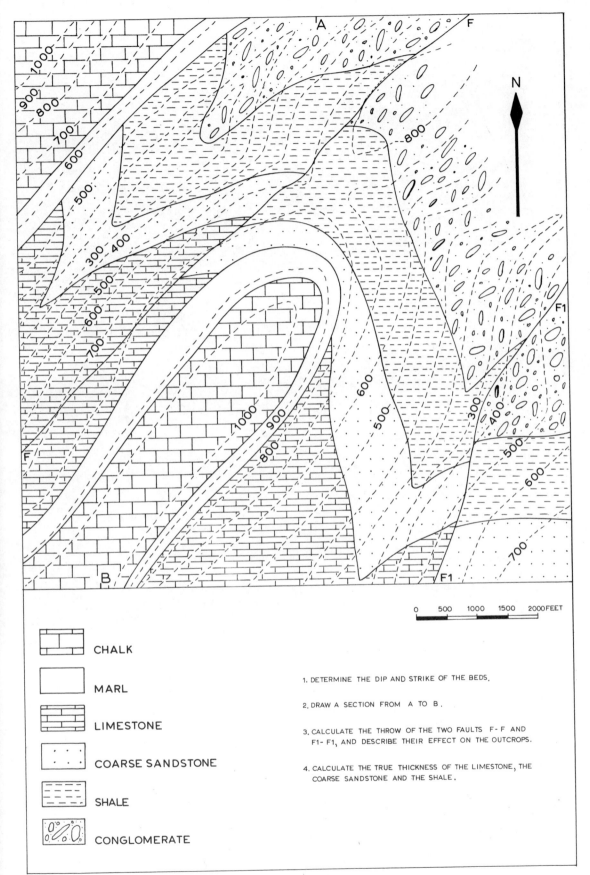

CHALK

MARL

LIMESTONE

COARSE SANDSTONE

SHALE

CONGLOMERATE

1. DETERMINE THE DIP AND STRIKE OF THE BEDS.

2. DRAW A SECTION FROM A TO B.

3. CALCULATE THE THROW OF THE TWO FAULTS F-F AND F1-F1, AND DESCRIBE THEIR EFFECT ON THE OUTCROPS.

4. CALCULATE THE TRUE THICKNESS OF THE LIMESTONE, THE COARSE SANDSTONE AND THE SHALE.

FIG. 85

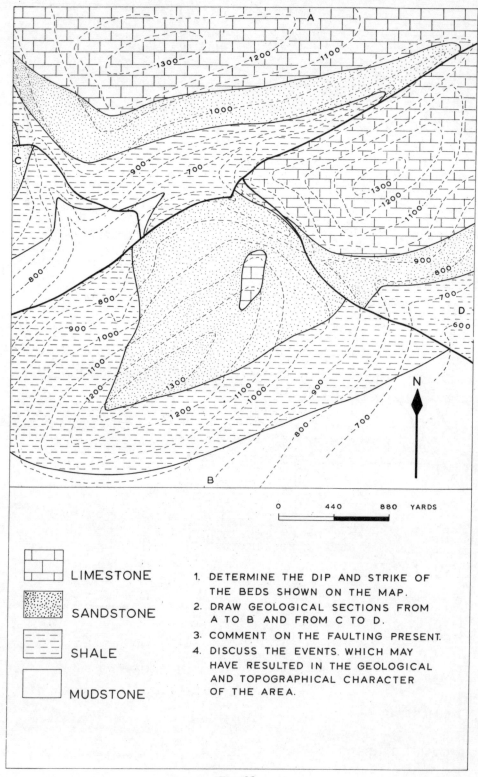

Fig. 86

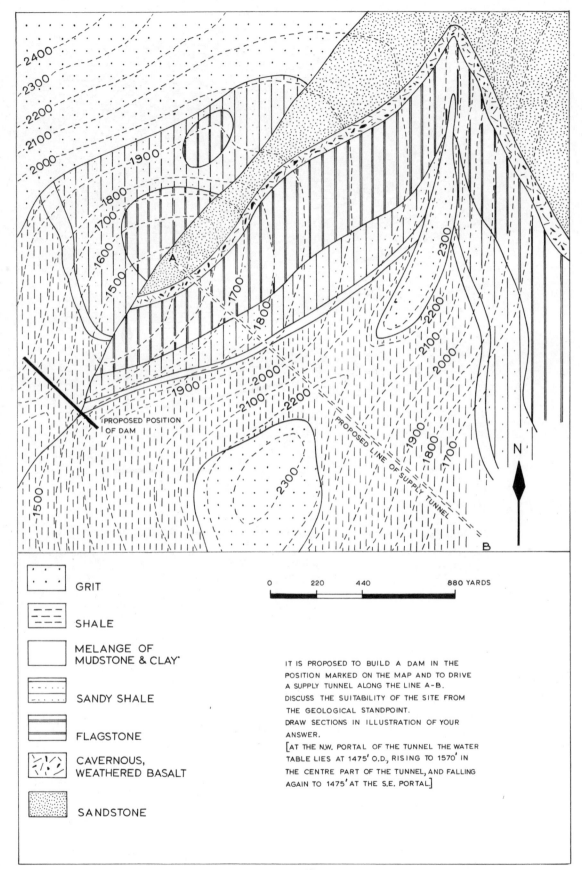

	GRIT
	SHALE
	MELANGE OF MUDSTONE & CLAY
	SANDY SHALE
	FLAGSTONE
	CAVERNOUS, WEATHERED BASALT
	SANDSTONE

IT IS PROPOSED TO BUILD A DAM IN THE POSITION MARKED ON THE MAP AND TO DRIVE A SUPPLY TUNNEL ALONG THE LINE A-B. DISCUSS THE SUITABILITY OF THE SITE FROM THE GEOLOGICAL STANDPOINT.
DRAW SECTIONS IN ILLUSTRATION OF YOUR ANSWER.
[AT THE N.W. PORTAL OF THE TUNNEL THE WATER TABLE LIES AT 1475' O.D., RISING TO 1570' IN THE CENTRE PART OF THE TUNNEL, AND FALLING AGAIN TO 1475' AT THE S.E. PORTAL.]

FIG. 87

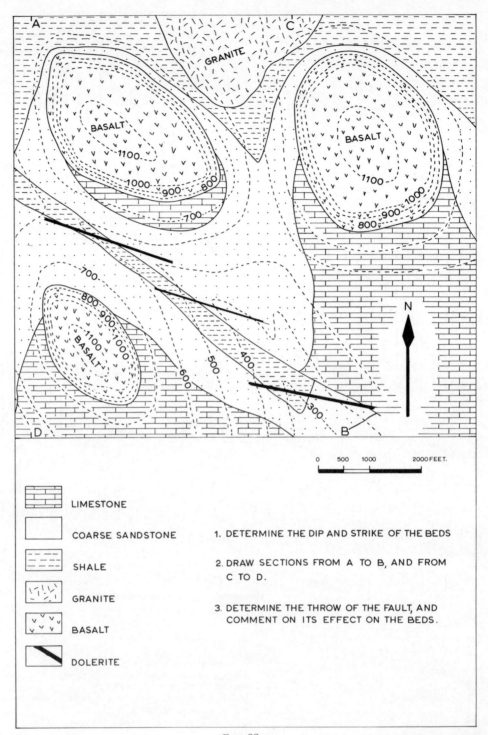

LIMESTONE

COARSE SANDSTONE

SHALE

GRANITE

BASALT

DOLERITE

1. DETERMINE THE DIP AND STRIKE OF THE BEDS

2. DRAW SECTIONS FROM A TO B, AND FROM C TO D.

3. DETERMINE THE THROW OF THE FAULT, AND COMMENT ON ITS EFFECT ON THE BEDS.

FIG. 88

BIBLIOGRAPHY

ANDERSON, E. M. *The Dynamics of Faulting*. Oliver & Boyd, 1942.

BENNISON, G. M. *Introduction to Geological Structures and Maps*. Edward Arnold Ltd., 1964.

BROWN, C. BARRINGTON and DEBENHAM, F. *Structure and Surface*. Edward Arnold Ltd., 1929.

CHALMERS, R. M. *Geological Maps*. Oxford University Press, 1926.

DWERRYHOUSE, A. R. *Geological and Topographical Maps*. Edward Arnold Ltd., 1942.

ELLES, GERTRUDE L. *The Study of Geological Maps*. Cambridge University Press, 1931.

HARKER, ALFRED. *Notes on Geological Map Reading*. W. Heffer & Sons, 1926.

NELSON, A. *Geological Maps–Their Study and Use*. Colliery Guardian Co. Ltd.

ROBERTS, A. *Geological Structures and Maps*. Cleaver-Hume Press Ltd., 1958.

SIMPSON, BRIAN. *Geological Map Exercises*. Geo. Philip & Son Ltd., 1960.

INDEX

African Rift Valley 32
Ammanford 32, 55
Anticline 16, 17, 18
  asymmetrical plunging 22
  axial plane of 16
  crestal plane of 16
  plunging 21
  symmetrical 16, 40
Assynt 51
Axial plane 16, 18, 20, 21
Axis of fold 21

Bath 51
Bed, thickness of 14, 15
Bedding planes 3
Borehole 7

Careg Cennen Castle 55
Chalk 7, 8
Conformable relationship 6
Contour line 1

Devizes 13
Dip 3, 5–7, 9–16, 18, 24
  angle of, in exaggerated vertical scale 68
  apparent 3, 71
  relation between true and apparent 75–76
  true 3, 69–72
Dyke 3

Escarpment 9

Faults, 26–48
  classification of 27
  descriptive terminology of 26
  determination of throw of 33, 41–44
  dip 29, 30, 34, 39–40, 42
  dip of 26, 27, 45, 46
  dip slip 28
  direction of slip of 27
  downthrow side of 26–27, 29, 34–41
  effects of, on outcrop of bed 33
  hade of 26–27, 30, 34–38, 45–46
  heave of 27
  hinge 31, 33
  low angle 48
  movement of 29–30
  normal 27, 29, 30, 36–37
  normal strike 36, 37, 41, 45–46
  oblique 29–30
  oblique slip 28
  reverse 29, 30, 34, 35, 38
  reverse dip 34
  reverse strike 35, 38, 41
  sag 31, 33
  stratigraphic throw of 27
  strike 29, 30, 35–38, 43–46
  strike slip 28
  swivel 33
  tear 31
  throw of 26–27, 33, 41–44
  trace of 36
  types of movement of 29
  upthrow side of 26–27, 29, 34–41
  vertical throw of 27
Faulting 34
  low angle 48
Fault plane 26, 27
  direction of movement on 29
Fold
  asymmetrical 16–17
  isoclinal 18, 20
  pitching 18
  plunging 18, 20–22
  recumbent 18–19, 25
  symmetrical 16–17
Frome 49

Geological Survey of Great Britain 1
Gower 33
Graben 32

Horst 32

Igneous rock 3
Inlier 55, 56
  faulted 55, 57

Leicester 49

Map
  Ordnance Survey 1
  topographic 1
Matterhorn 18
Metamorphic rock 3
Midland Valley of Scotland 33
Miscellaneous examples 82–95
Mount Sorrel Granite 49

Outcrop 3, 5, 13
  completion of, from partial outcrops 58, 59, 62
  plotting of, from borehole records 58, 60–61, 63–67
Outlier 50, 55–56
  faulted 55, 57
Overfold 18–19
Overlap 51
Overstep 51

Plateau surface 9

Reigate 55
Rift valley 32

Sanquhar 49
Sedimentary rock 3
Sill 3
Slip 27, 29
  dip 27, 29
  oblique 27, 29
  strike 27, 29

Southern Uplands 18
South Wales Coalfield 33
Stereographic projection, uses of, in solution of dip and strike problems 77–81
Strike 3
Strike line 3, 5–7, 9–14, 16, 18, 22, 24
Syncline 16–18
  asymmetrical plunging 22
  axial plane of 16
  faulted symmetrical 39
  plunging 21
  symmetrical 16, 39
  trough of 16

Thickness of a bed 14, 15
"Three-point-problems" 58
Topography 5, 13

Unconformable relationship 52, 54
Unconformities 49, 51, 54
  plane of 50, 51
Uniclinal strata 16
Usk 55

Vale of Clwyd 33
Vallis Vale 49
Vosges 32

Wolff net 77–81

Zenithal point 77